Cambridge Elements

Elements in the Philosophy of Science
edited by
Jacob Stegenga
NTU Singapore

ANALOGICAL REASONING IN SCIENCE

Francesco Nappo
Catholic University of Milan
Giovanni Valente
Polytechnic University of Milan

CAMBRIDGE
UNIVERSITY PRESS

CAMBRIDGE
UNIVERSITY PRESS

Shaftesbury Road, Cambridge CB2 8EA, United Kingdom

One Liberty Plaza, 20th Floor, New York, NY 10006, USA

477 Williamstown Road, Port Melbourne, VIC 3207, Australia

314–321, 3rd Floor, Plot 3, Splendor Forum, Jasola District Centre, New Delhi – 110025, India

Cambridge University Press is part of Cambridge University Press & Assessment, a department of the University of Cambridge.

We share the University's mission to contribute to society through the pursuit of education, learning and research at the highest international levels of excellence.

www.cambridge.org
Information on this title: www.cambridge.org/9781009526463

DOI: 10.1017/9781009526494

When citing this work, please include a reference to the DOI 10.1017/9781009526494

First published 2026

A catalogue record for this publication is available from the British Library

A Cataloging-in-Publication data record for this Element is available from the Library of Congress

ISBN 978-1-009-52646-3 Hardback
ISBN 978-1-009-52647-0 Paperback
ISSN 2517-7273 (online)
ISSN 2517-7265 (print)

Analogical Reasoning in Science

Elements in the Philosophy of Science

DOI: 10.1017/9781009526494
First published online: June 2026

Francesco Nappo
Catholic University of Milan

Giovanni Valente
Polytechnic University of Milan

Author for correspondence: Francesco Nappo, francesco.nappo@polimi.it

Abstract: When faced with a difficult problem or limited information about a novel domain, how do scientists advance their research? As historians of science have widely noted, one strategy common to the natural and the social sciences is to make use of analogy. Formulating hypotheses about an unknown system construed by analogy with what is observed in a more familiar system has repeatedly proven to be a source of discoveries. But what makes analogy such a useful tool for scientific inquiry? Although early reflections trace back to Aristotle, the question of the exact role of analogy in science remains an outstanding one in contemporary philosophy of science. This Element aims to clarify the main epistemological questions at stake and why seemingly obvious answers to them do not survive scrutiny. We provide an overview of the current debate and summarize insights from relevant case studies in the natural and social sciences.

Keywords: analogy, surrogative reasoning, scientific modeling, inference, experimentation

ISBNs: 9781009526463 (HB), 9781009526470 (PB), 9781009526494 (OC)
ISSNs: 2517-7273 (online), 2517-7265 (print)

Contents

Introduction

Mathematician George Pólya is often credited with the sentence: "if you can't solve a problem, there is an easier problem you can solve: find it" (cf. Polya 1957:114). His advice reminds us that the capacity to reconceptualize a complex problem in more familiar terms is a crucial element of human cognition. Many breakthroughs, in research as well as in everyday life, are achieved in precisely this way. When confronted with a problem that resists solution, we search our memory for hints that could allow us to circumvent the obstacle. Frequently, these recollections lead nowhere. Sometimes, however, what initially appeared as a vague resonance with another problem or domain turns out to reflect a deeper resemblance. We are then led to view the original problem as tractable via familiar tools, gaining confidence as further correspondences emerge.

Seeing known patterns in what is unfamiliar by identifying matching features is characteristic of analogical reasoning. The strategy of retrieving an analogue as a tool for problem solving occurs in virtually all fields of scientific research, ranging from physics, biology, and chemistry to economics, medicine, and archeology (Hesse 1966). Its role during the exploratory stages of scientific investigation emerges clearly from in vitro studies of research environments (Dunbar and Blanchette 2001) as well as from the testimony of working scientists (e.g., Oppenheimer 1956). Albeit frequently a source of false expectations and inadequate theories, analogies also figure prominently in some of the most significant breakthroughs in the history of science, from Galilei's discoveries of lunar valleys and mountains to Darwin's theory of natural selection.

Philosophically, the use of analogies in scientific research presents a puzzle. Insofar as they depend on subjective judgments of similarity, we would expect analogies to mostly serve functions related to science communication and education. How can they instead also shape scientific investigation into the unknown, to the point that entire research programs are sometimes designed and pursued on their basis? Our aim for this Element is to explain in an accessible way the main epistemological issues at stake and the different positions defended on this subject. To assist nonspecialist readers in navigating a scattered literature, we provide a systematic account of the current debate and summarize insights from relevant case studies. Ideally, the way we structure the existing discussion will allow for an in-depth assessment of the arguments on different sides. As we will see, philosophical questions about the definition and role of analogy in science are deeply intertwined with core issues in scientific methodology, including the validity of experimental results beyond laboratory settings, the limits of empirical confirmation, the unity or disunity of the sciences, the design and responsible use of models in decision-making, and many others. An account of scientific methodology cannot dispense with addressing the foundational questions raised in the following sections. By summarizing the main alternatives and points of disagreement, we aim to stimulate further philosophical reflection on these important topics.

For reasons of space, this Element will not cover the most technical parts of the subject, including different methods for formalizing analogical reasoning through logico-mathematical structures (Carnap 1980; Skyrms 1993) and neural network architectures (Mitchell 2021). As current artificial systems are said to fail the test of so-called "general" intelligence, viz., of emulating the human mind's plasticity in adapting to problems of different kinds, formal work on analogy has recently become a fast-growing research area (Holyoak 2025). While we will not attempt a summary of key findings, readers who are interested in recent formal approaches will still find some utility to this volume. Analyzing skillful uses of analogy in scientific contexts helps review the presuppositions about analogical reasoning that lie behind formal approaches, providing insights about their strengths and potential pitfalls in emulating the workings of a human mind.

For similar reasons, the contents of this volume should also be relevant to experts in science education. Among the most critical areas of interest in this discipline is problem-solving skills (Berieter 1985). The hardships faced by students in framing problems correctly, identifying solution strategies, implementing those strategies independently, and reflecting on the results are well-documented (Niss and Blum 2020). While developing a capacity for analogical reasoning is often viewed as crucial for activating creative thinking, different accounts of analogy suggest different ways to foster it and different benchmarks for testing it. By synthesizing insights from several disciplines, a philosophical perspective can help ensure that educational theory and practice align with a correct understanding of the sought-for problem solving capacity.

The discussion below is organized as follows. Section 1 introduces the puzzle that this volume zeroes in on: the role of analogy at the early stages of scientific research. After clarifying some basic distinctions between "analogy," "analogical reasoning," and "analogical inference," we outline a general framework for classifying different conceptions of the notion of analogy in science, based on the type of similarities and method of aggregation. We then distinguish two broad perspectives on the epistemological puzzle raised by their role at the early stages of scientific research. The *heuristic view* emphasizes the contribution of analogy to creative thinking and the generation of novel ideas, whereas the *inductive view* holds that analogy can also play a role in the justification of empirical hypotheses before they are subject to testing. The ambiguity surrounding the notion of analogy, together with the existence of different perspectives concerning its role in empirical research, provides the motivation for developing a philosophical theory.

Section 2 reviews attempts in cognitive science and philosophy to provide general criteria of plausibility for analogical inference in science, with the aim of understanding whether there are some precise conditions under which scientists form expectations about the truth or falsity of empirical conclusions based on analogy. This question is particularly important for defenders of the inductive view, insofar as they are committed to explaining when and how analogy can do justificatory work. After surveying several proposals, however, Section 2 concludes that no existing theory satisfies all desiderata. Particularly noteworthy in this discussion are the criticisms advanced to influential theories of plausible analogical inference in contemporary cognitive science, starting from the so-called "structure-mapping" theory by Dedre Gentner and colleagues, given their current impact on both research in artificial intelligence and in science education.

Section 3 considers the perspective of scientific practice, asking to what extent we can expect a normative theory of analogical inference to extend to the variety of contexts in which analogy is seemingly at work. Our main contribution in this section – by far the most technical and specialistic – consists in bringing attention to an underappreciated problem for the inductive view about analogy, which is that of determining the boundaries within which normative criteria apply to the inferential practices of scientists. We show how this issue, that we call the *problem of extendibility*, lies behind several current debates in contemporary philosophy of science, such as those concerning the epistemology of analogue simulations in physics, the methods of extrapolation in the biomedical and social sciences, the interpretation of the phenomenon of scientific model transfer, and the evaluation of predictive data analysis tools based on formal similarity metrics.

Having noted how the prospects for the inductive view vis-à-vis the heuristic alternative are never entirely foreclosed, Section 4 concludes by reviewing candidate answers to the problem of what could justify scientists in forming expectations about the unknown on the basis of an analogy. The main solution sketches are based on principles of symmetry or, alternatively, of simplicity.

1 Analogy in Scientific Investigation

Let us start from a classic example of the use of analogy in scientific research: The development of the Rutherford–Bohr model of the atom (Rutherford 1911; Bohr 1913):

E1. The atom as a Solar system

The hypothesis that atomic structure may be similar to that of the Solar system had already appeared within the physics community decades before the discovery of the electron. In a model elaborated by Larmor in 1897, a positively charged nucleus is surrounded by negatively charged masses moving along orbits, where mutual attraction is generated by the electrostatic force rather than gravity. Early analogies with the Solar systems were, however, plagued by an instability problem. According to classical electrodynamics, rotating charges should emit radiation and hence quickly collapse into the nucleus as they lose energy, unlike what happens in the Solar system. In 1911, though, experiments by Rutherford on scattering of alpha particles reinforced the idea that the atom should have a highly massed nucleus at the center. Inspired by the analogy between the atomic nucleus and the Sun, Niels Bohr set himself to resolve the instability problem by introducing Planck's hypothesis that the energy emitted by radiation is quantized, so that electrically charged masses move around stable orbits of discrete energy levels and radiate only when jumping from one orbit to another of lower energy (Bohr 1913). The transition to quantum electrodynamics also enabled Bohr to recover some other (yet unexplained) observations, such as the discrete spectrum of the hydrogen atom, thereby constructing the first physically consistent atomic model.

This example highlights some key elements characterizing analogical reasoning at the early stages of research. To begin with, when the Rutherford–Bohr model was constructed, empirical access to the atomic scale was severely limited, prompting the search for suitable analogues. A potential connection with the workings of the Solar system could be suggested by mathematical similarities, such as that electrostatic force obeys the same inverse square law as gravitational attraction, as well as by empirical facts, including Rutherford's demonstration that the mass of the atom is concentrated in the nucleus just like the mass of the solar system is concentrated in the Sun. Depicting the atom along the lines of a more familiar structure offered a tractable picture of an otherwise unknown system. Indeed, even though Bohr's model was later superseded by more sophisticated theories, the analogy with the Solar system is still employed today in physics education for its intuitiveness. At least as important as its cognitive tractability, however, is the role of the analogy as a basis for articulating *inferences* about the target system. By suggesting that electrons orbit around their nuclei in familiar ways, the model provided both a provisional explanation of the available experimental observations and the means to devise further tests that could either confirm or disconfirm it. The process that goes from the stage of retrieval of an analogue to the mapping of the similarities and eventually to the articulation and evaluation of inferences about the target domain is what, in this Element, we refer to as *analogical reasoning*. It is importantly distinct from *analogical inference,* which refers to the specific process whereby known similarities between source and target are used as a basis to infer some further similarity. An *analogical argument* is a linguistic representation of this inferential stage of analogical reasoning.

Another element that emerges from studying the Rutherford–Bohr model concerns the aspect of selection of the analogue and the assessment of the similarities. The Solar system is identified as an analog of an atom based on their perceived similarities, notwithstanding the fact that the two systems differ in many important respects. For instance, the atom is a microscopic object while the Solar system is astronomically large; moreover, attractive forces act between charges instead of masses. This illustrates that a central capacity related to the selection of analogues is the ability to see commonalities, if not among the objects themselves, in the *relations* that they entertain with one another – such as the relation of attraction that links a planet to the Sun. At the same time, the decision to use a given source as an analogue requires a choice as to how much dissimilarity can be tolerated. A close analysis of various episodes in the history of science reveals that there is no univocal attitude among researchers toward the recognition of *dissimilarities*: Sometimes they prompt a scientist to divert attention from an alleged analogy, as in the case of the early work on the planetary model until Rutherford's results revived it, while other times they are ignored for the sake of making indirect inferences about the target, as in the case of Bohr. Hence, the assessment of an analogy is complex, context-sensitive, and inherently disputable. These aspects should figure in any satisfactory account of analogical reasoning in science.

When assessing the role of analogy in the scientific context, it is helpful to distinguish between different types of issues. At a *cognitive* level, there are issues concerning how it happens that scientists interested in studying a system come to see similarities with another distinct system and how they go on to infer that there might be other salient similarities. Understanding the mechanism behind individual and group perceptions of resemblance, inference, and belief formation pertains to the realm of psychology. These issues, albeit interesting to explore, mostly fall outside the scope of the present Element (see Holyoak 2025 for a recent treatment). More aligned with our purposes is the *descriptive* level of assessment of analogical reasoning, wherein one tries to reconstruct the inferential practice adopted in science. Describing how scientists use, or have used, analogies with more familiar domains gives insight into the various uses of analogy. The central interest of this Element, though, concerns the *normative* level of assessment, wherein one tries to identify the conditions under which scientists are justified in their judgments of similarity and in their drawing analogical inferences. As discussed in Section 1.2, the hardest epistemological problems regarding the role of analogy in science concern this third level.

In the remaining part of this section, we proceed to unpack the key philosophical issues about analogy. In Section 1.1, we address the issue of what exactly should be regarded as an analogue system, also distinguishing different notions of "analogy" that are present in the literature. Next, in Section 1.2, we set the stage for an in-depth treatment of the problem of justification for analogical inference in science by outlining the two competing views, the *heuristic* and the *inductive* view.

1.1 What Is an Analogue?

Reasoning by analogy is a form of "surrogative reasoning." Used by Swoyer (1991) to denote the process of drawing inferences about a target system from a theoretical model, this term can be generally extended to encompass the practice, ubiquitous among scientists facing epistemic limitations about a system under study, to resort to a surrogate system to gain some indirect information. But while analogical reasoning fits this characterization, not all reasoning from surrogates can be considered analogical. The practice of building *models of data* (Suppes 1962), for instance, is not easily interpretable in terms of analogy, even though it yields surrogates of a sort. Models of data are commonly used to clean or rearrange raw data and are assessed mainly based on their fit with the observations, rather than on some form of similarity between a source and a target system (Bayler-Jones 2002). A borderline case is given by *scale models*, viz., copies of a system at more accessible scales, which count as analogues on some definitions but not others (Sterrett 2020; Suarez 2024). A question that we must therefore pose is the following: Out of all species of surrogative reasoning, what is distinctive about analogical reasoning? In particular, what makes a given system an *analogue*? And what makes an inference *analogical*?

Ordinarily, we say that two systems (which for our purposes may be material objects, domains of discourse, or even abstract or fictional entities) are analogous when they are *similar* in significant respects. Sharpening this definition for application to the scientific domain requires specifying the relation of similarity to adopt, as well as which properties should be regarded as relevant and which types of entities the analogical relation can hold. There is a vast array of proposals in both the scientific and the philosophical literature concerning these definitions. In scientific works, a recurrent trend is the attempt to determine a restrictive meaning of "analogy" in a particular discipline, such as physics (see, e.g., Maxwell 1890; Mach 1905; Oppenheimer 1956) or biology (Lorentz 1974), with the aim to show that instances of the definition fulfil a specific function. In the philosophical literature, substantive work has been done on the notion of similarity (Goodman 1972), most recently in the context of the epistemology of scientific models (Weisberg 2013; Suarez 2024). Among other things, philosophers debate whether all scientific models represent their targets in virtue of being similar to them (Suarez 2003; Giere 2004; Frigg and Nguyen 2020). Our aim in what follows is to provide some form of simplification of this highly heterogeneous field of proposals by means of a typology of the main conceptions of the notion of analogy in science.

Before we proceed in the classification work, a prior obstacle must be taken care of, which is that not all systems that may be judged as similar to a given target count as analogues. A plausible example comes from *model organisms*, which are biological organisms with artificially selected traits that are used as research materials in biology and medicine. In Ankeny and Leonelli's (2020) terms, these organisms differ from mere analogues in having broader *representational scope*. That is, whereas analogue systems are used to draw inferences about specific targets, model organisms such as *Drosophila melanogaster* are used to draw generalizations about the whole class of processes and mechanisms of which the models are instances, based on features such as their simplicity and shared ancestry (Levy and Currie 2015). Strictly speaking, then, inferences from model organisms are not analogical. What this suggests is that an important conceptual requirement on any notion of an analogue system is *narrow* representational scope, that is, its being used as a *surrogate for* a specific distinct system – as in the case of the analogy between atom and Solar system (Section 1, E1) – rather than as specific *instances of* a general class of entities.

As anticipated, even within the class of surrogates possessing the right representational scope, there are many proposals in the literature regarding how similarity is established and used as a basis for reasoning. To achieve simplification, here we will classify the conceptions based on two factors: *type of similarities* and *aggregation method*. Depending on the combination one favors (either in general or in a specific field), different systems may, or may fail to, count as analogues.

1.1.1 Type of Similarities

Similarity is not tantamount to strict identity. This renders the notion of similarity more liberal and yet less prone to be pinned down rigorously. At least two conceptions can be distinguished:

Within-type – Similarity can be intended as a relation between token properties or relations of the same type. Early versions of the structure-mapping theory by Gentner (1983) incorporate a requirement (later abandoned) that similarities and dissimilarities are with respect to identical predicates (see Section 2.1). For instance, after Rutherford's 1911 experiments suggesting that the nucleus is considerably more massive than electrons orbiting around it, the analogy with the relative masses of the Sun and its orbiting planets contributed to reviving the planetary model of the atom. Because a fact about the distribution of mass is common to both source and target systems, here we may speak of *within-type* similarity. When similarity is conceived as within-type, it is possible to make direct comparisons between target and source with regard to the relevant properties, for instance, by checking whether there is any matching between respective values.

The notion of within-type similarity admits of degrees. In a *perfect* within-type similarity, the values that are projected from the analogy's source yield (possibly *modulo* some fixed transformation) exactly the target properties' known values. Of course, most often we deal with *imperfect* cases. When the values that are projected from source to target are acceptable approximations, the fact that they instantiate lesser degrees of within-type similarity does not affect the analogy. For instance, in the Rutherford–Bohr model, the relative mass distribution of nuclei compared to electrons need not be exactly proportional to the relative mass distribution of the Sun compared to the planets in order for the planetary model to be plausible. For an approximation of the target properties to be good, or even just acceptable, one needs to assume some margins of tolerance. This has the drawback of introducing an element of arbitrariness within-type similarity.

Cross-type – Alternatively, similarity can be defined across properties or relations that are not of the same type. Many philosophical and cognitive science accounts of analogical reasoning, including later versions of Gentner's structure-mapping theory, countenance cross-type similarity as admissible for plausible analogical inference. In cross-type similarity, a direct quantitative comparison among the relevant properties may not be a viable option. However, one can admit to such similarities by judging the overall behavior or role that certain properties or relations play in the target system and establishing some correspondence with those of the source. In the Rutherford–Bohr model, for instance, the gravitational force between the nucleus and electrons is too weak to explain the stability of the system; this stability is mainly due to the electromagnetic force that attracts the opposite charges of the protons in the nucleus and the electrons. The similarity between gravitational and electrostatic forces can therefore be categorized as *cross-type*.

Due to its appeal to similar behavior or similar roles, the notion of cross-type similarity can relate systems from very different contexts. It is customary to distinguish resulting analogies in their degree of *closeness*, depending on contextual factors. For example, that between gravity and electrostatics can be classified as a relatively close analogy, in contrast to distant comparisons between thermodynamical phase transitions and financial crashes as studied in the field of econophysics (cf. Palacios and Jhun 2023). Compared to within-type, cross-type similarity introduces further dimensions of arbitrariness in the evaluation. For one, given that the corresponding properties are not of the same type, there can hardly be quantitative methods to assess, even just approximately, how close their respective values are. Moreover, since judgments of cross-type similarity exploit aspects related to functional role or behavior, it becomes harder to adjudicate which properties are relevant across the different contexts, as well as to justify why certain pairs of properties rather than others should be viewed as corresponding.

1.1.2 Aggregation Method

A conceptually distinct issue is how similarities and dissimilarities are assessed collectively in a judgment of analogy. After all, for an analogy to be drawn between a source and a target, one typically needs more than one similarity; at a minimum, one would expect a whole series of them. We can distinguish three approaches to similarity aggregation: *relevance-driven*, *structural*, and *geometric*. Below and throughout the Element, we will conventionally denote the target's properties with unstarred symbols $p_1, \ldots, p_n$ and the source's properties with starred symbols $p_1^*, \ldots, p_n^*$.

Relevance-driven approaches – One approach to the aggregation of similarities – close to the ordinary way of assessing analogy – conceives of the evaluation of an analogy as dependent upon how many properties $p_1, \ldots, p_n$ of the target are relevantly similar to the respective properties $p_1^*, \ldots, p_n^*$ of the source (so-called *positive analogies*) and how so (i.e., to what degree), as well as how many properties are relevantly dissimilar (so-called *negative analogies*) and how so, where the degree of relevance is viewed as a function of the analogy's intended use or purpose. This is the approach to similarity aggregation suggested in Hesse's (1966) influential treatment of analogical reasoning and successive developments (e.g., Gentner's 1989 *systematicity criterion* and Bartha's requirement of *prior association* discussed in Section 2). A relevance-driven approach describes what often happens in scientific practice, where the more relevant similarities scientists see, the more they may be inclined to exploit the analogy to investigate a target. From a normative perspective, however, the approach has several shortcomings. For one, it does not define just how large the set of positive analogies should be in order for the source to be an analogue of the target. Moreover, it is not clear whether any rule, or even vague recommendation, can be articulated regarding how to proceed in the face of negative analogies between a source and a target, or in the presence of multiple candidate analogues. Defenders of relevance-driven methods typically deny that such matters can be determined without specifying a context.

Structural approaches – Another approach to aggregation assesses analogy based solely upon formal features of the *relations* holding between corresponding properties of source and target. In general, we say that there is an *isomorphism* between two sets of properties $p_1^*, \ldots, p_n^*$ and $p_1, \ldots, p_n$ when there exists a one-to-one mapping that associates each property in one set with exactly one property in the other, and vice versa, regardless of whether $p_1^*, \ldots, p_n^*$ and $p_1, \ldots, p_n$ are properties of the same type. Formal desiderata may then be placed over the isomorphism to evaluate the overall degree of analogy. For example, a candidate formal desideratum (incorporated in Gentner et al.'s "structure-mapping theory" discussed in Section 2.1) is the criterion of *parallel connectivity*, meaning that if two relations are in one-to-one correspondence, so must their relata individually. Although the application of formal desiderata has the virtue of making the notion of correspondence among properties precise, thereby also yielding a measure of how overall "similar" two systems are, it often fails to yield a single analogue. Even when a common structure is identified that satisfies all formal desiderata, selecting a single "best" analogue may still prove impossible in situations of isomorphism with multiple candidate sources.

Geometric approaches – A third approach to aggregation resorts to formal similarity metrics defined over sets of properties of source and target, which yield a measure of how overall similar two objects or systems are. With the development of big data, techniques routinely applied to data analysis avail themselves of such metrics. Similarity analyses are typically used to classify a given population into clusters, by assigning individuals or groups under study to their overall most "similar" cluster (Boniolo et al. 2023b) based on a formal *similarity metric*. The sets of properties over which the metrics are defined can be identified either conventionally, based on arbitrary sets of available data, or evidentially, based on a prior selection of the properties that are considered most relevant for a given use. In the geometrical approach, one compares source and target with respect to all the available properties $\mathrm{p}_1^*, \ldots, \mathrm{p}_\mathrm{n}^*$ and $\mathrm{p}_1, \ldots, \mathrm{p}_\mathrm{n}$, but there does not need to be one-to-one similarity across them. In most practical applications, similarity is intended in the *within-type* sense; that is, the properties p_k^* and p_k are taken to be tokens of the same type. Formal approaches such as the *conceptual spaces* framework (Gãrdenfors 2000) may be used to extend geometric aggregation methods to cases of cross-type similarity (Douven et al. 2025). An advantage of the geometric approach is that questions about how much similarity is required and how much dissimilarity is tolerable for a given analogical inference can be answered precisely for a given application. At the same time, there are different kinds of similarity metrics one can adopt, resulting in different "most similar" analogues: while typically aggregation occurs by averaging the values of all relevant properties for each system, or the differences between the corresponding properties, definitions based on more complex topologies are possible and useful in some cases (Grenier et al. 2013). Also, compared to intuitive judgments of analogy, geometric aggregation tends to be somewhat insensitive to individual similarities or dissimilarities. Consequently, it may happen that an analogue system may score highest based on a metric even though it presents what scientific experts would ordinarily consider critical dissimilarity with the target – one that impedes plausible analogical inference. We will return to scientific applications in Section 3.4.

Unlike the distinction between within-type and cross-type similarity, it is worth stressing that distinct aggregation methods do not mutually exclude one another: They are merely different ways of establishing a sense in which target and source are "analogous," which in turn may depend on the context of application and purpose of the investigation. As will be discussed in Section 2, normative theories for analogical reasoning in science sometimes combine distinct methods. For instance, Gentner's (1989) influential structure-mapping theory uses structural and relevance-driven approaches in a sequential way to identify the "best" analogue for a given source.

1.2 Analogical Inference

As discussed in Section 1.1, analogical reasoning is a complex type of reasoning. Cognitive scientists distinguish the stages of retrieval of an analogue, the mapping of the known properties of the target to those of the retrieved analogue, and the inference to the corresponding properties of the target (Gentner 2010; Holyoak 2025). Having surveyed various conceptions of what can be in an analogy, viz., which types of analogues can be retrieved and mapped in a context of scientific research, let us now turn to discuss analogical inference. As anticipated, this stage raises the hardest questions from an epistemological viewpoint. Conventionally, the inferential stage is represented by the following *argument schema*, where T stands for target and S for source:

Analogical argument schema:

T and S are similar with regard to properties $p_1, \ldots, p_n$ and $p_1^*, \ldots, p_n^*$, respectively.

T and S are not dissimilar with regard to critical properties $p_k, \ldots, p_m$ and $p_k^*, \ldots, p_m^*$,

S possesses property p_{n+1}^*

Therefore, T possesses the corresponding property p_{n+1}

The first premise in this argument schema constitutes what we can call the *similarity basis* of the inference, which sets the conditions for a given source to be an analogue of the target of interest. As we have seen, the identification of this basis depends on the specific notion of similarity one adopts (within-type or cross-type) as well as on the preferred approach to the aggregation of similarities. For instance, in an argument to the effect that electronic orbits are stable in analogy with the stability of planetary orbits (Section 1, E1), the similarity basis is given by the correspondences between the relative mass of the nucleus and the electrons, on one hand, and the relative mass of the Sun and the planets, on the other. The similarity basis is the *driver* of an analogical inference.

The second premise in the argument schema states what we may call an *enabling clause* for the analogical inference: in order for the conclusion to be plausible, it must be the case that the target lacks *critical properties* $p_k, \ldots, p_m$. By "critical," we mean that their possession is judged, based on background knowledge, to prevent the target from possessing p_{n+1}. Trivially, if $\neg p_{n+1}$ is the case – for instance, if we knew electronic orbits are not stable – the analogical inference is not licensed. However, a broader range of properties are typically classified among the "critical" ones, with the aim of ruling out analogical inferences that are eminently weak or risky. For instance, in assessing the inference to the stability of electronic orbits (Section 1, E1), one may note that electrostatic forces in the atom present different characteristics from gravitational forces occurring between the Sun and planets. The decision concerning which properties are critical relative to an inference can be based on evidential or pragmatic considerations, or both (Hesse 1966: 89; Bartha 2009: 25).

Assuming that both similarity basis and enabling clause are in place, reasoning from those premises to an intended conclusion acquires *inferential force* just if the additional property p^*_{n+1} of the analogue and the corresponding property p_{n+1} attributed to the target can be *related* to the sets of similar properties $p^*_1, \ldots, p^*_n$ and $p_1, \ldots, p_n$, respectively. The term "corresponding" signals that, if it obtains, the property p_{n+1} that is inferred must bear the same (or similar) relation as the one that relates $p^*_1, \ldots, p^*_n$ to p^*_{n+1} in the source. For example, we expect the property of stability in electronic orbits to derive from a similar configuration of forces in the atom as the one that relates the Sun to the planets of the Solar system, which is the cause of the stability in planetary orbits. The evaluation of inferential force extends the method of similarity identification and aggregation employed in identifying the analogue across potential source systems. Hence, on a structural approach, the similar properties $p_1^*, \ldots, p_n^*$ and $p_1, \ldots, p_n$ are selected based on the existence of an isomorphism relating them, as in the structural correspondence between properties of atoms and properties of the Solar system (Section 1, E1); for the sake of analogical inference, this isomorphism must formally extend to another one-to-one mapping between the properties p_{n+1}^* and p_{n+1}, which figure in the conclusion of the analogical inference (e.g., stability of orbits). In relevance-driven and geometric approaches, instead, inferential force depends on extending the same type of relations among properties used in evaluating the relevance of the similarity basis; these may range from purely conventional and statistical relations to causal and explanatory connections.

Inferences that satisfy the argument scheme emerge from many different contexts of scientific research, raising a common epistemological problem. Since source and target systems are supposed to be nonidentical, at some point their alleged analogy should break – that is, they should differ with respect to some property. Indeed, even when the two systems share many properties, or even all known properties, it does not logically follow that, for any additional property observed in the source, the target must have a similar property. Granted that the conclusion is uncertain, and that sometimes it cannot even be verified or falsified *a posteriori*, one faces the following problem: Under what conditions is drawing an analogical inference justified in science? This problem applies to many different contexts, including cases where the interpretation of experimental results is at stake or where the suitability of a research method is under scrutiny (Stegenga 2024). However, a particularly prominent instance – long emphasized by philosophers of science concerned with analogy in science – centers on situations where the object of inquiry is whether a given hypothesis about a system of interest holds. Before turning to outlining the competing views on this subject (in Section 1.2.2), one remaining clarification is necessary.

1.2.1 Epistemically Impoverished Contexts

The problem of justification arises when our knowledge about a target system is limited. In these cases – what we refer to as *evidentially impoverished* contexts –, the scarcity or sparseness of the information that we possess about the target makes it necessary to use indirect sources to acquire information. Evidently, there is no problem of justification when the truth or falsity of a conclusion could already be deduced from background information. For instance, a textbook case of analogy in physics is the so-called "Reynolds analogy" (Norton 2021). It expresses the empirically well-confirmed fact that, up to a certain level of approximation, transport phenomena concerning momentum, matter, and heat are physically identical. The fact that, modulo some fixed transformation rule, we can translate facts about losses of momentum in a fluid in a source into predictions about heat losses in a target, and vice versa, poses no special epistemological problem. Since the empirical generalization that grounds the analogy is independently well-confirmed, the similarity between the distinct physical systems is not the driver of the inference.

The sense in which we say that a context is "evidentially impoverished" should be understood in a relative manner. For one, it is a matter of degree. At one extreme, the analogue system may be regarded as a richer source of information, in that its properties are generally better known than those of the target; or it is empirically more accessible, by allowing for direct observations and experiments, which are less feasible to conduct on the target system. At the other extreme, there are cases in which there can be no empirical access at all to the target, even in principle. An example is black holes – regions of space-time with a gravitational pull so strong that no light can escape from within. In such extreme cases, important epistemological questions arise as to whether any hypothesis can be formulated and even supported by analogy (Crowther et al. 2019).

In addition to the degree of empirical access, we may classify contexts of research based on the reasons why they are evidentially impoverished. In this respect, we can distinguish several types of limitations on the accessibility of the target (where sometimes a context is evidentially impoverished due to multiple limitations). An important question that we can then ask is whether some of these limitations or combinations thereof pose an in principle obstacle to the use of analogy or to its ability to justify (to some extent) conclusions about an unknown target. We will return to this question in Section 4. For the moment, let us list the main limitations:

Instrumental – Our observational means and ability to make experiments about the target system can be constrained due to technological limitations. The history of particle physics contains plenty of examples. For instance, when the hypothesis of the Higgs mechanism was formulated, no available instrumentation could detect physical effects at such a low scale. To formulate the hypothetical mechanism for the generation of mass for gauge bosons, Higgs (1964) elaborated an analogy with the phenomenon of spontaneous symmetry breaking in condensed matter physics, specifically with familiar models for superconductivity (Fraser and Koberinski 2016).

Temporal – The target system may not be presently available. It could be a system from the past, for which we only have a few traces left. For example, a common exercise in archaeology is to reconstruct the function of an ancient remains based on similarities with other remains whose function is better understood (Wylie 1985; Currie 2018; Currie and Killin 2019). Alternatively, the target could be a system in the future that we compare to something present. For example, the methodology of spatial climate analogues aims to provide estimates of future temperatures and climate risk exposure for urban environments (Hallegatte et al. 2007; Valente et al. 2024).

Ethical – There can be moral constraints on performing experiments on the target. It then becomes appropriate to study some analogue for which certain ethical limitations do not arise, at least not as strongly. For example, experimentation on human beings is often disallowed in medicine, especially when the effects of trials can be dangerous or invasive. Experimentation on nonhuman animals that are known to be genetically close to human beings is more permissible, although moral constraints apply in those cases too (LaFoyette and Shanks 1993; Steel 2008).

Financial – Sometimes experimenting on a target is costly. Exploiting analogues can make research more affordable. For example, rhodium, a chemical element with atomic number forty-five in the platinum group of the periodic table, is one of the most expensive metals on Earth. Rather than carrying out experiments on it, it is more convenient to infer its properties from other elements either in the same row (similar physical properties) or the same column (similar reactions) of the table.

Theoretical – Sometimes a theory about the target system may be available, but scientists have reasons to regard it as incomplete or inadequate. For instance, certain aspects of the behavior of biological organisms and financial markets are not theoretically well understood. This has inspired the transfer of established physical models to biology and economics (Knuuttila and Loettgers 2016; Palacios and Jhun 2023). The Johansen-Ledois-Sornette (JLS) model, for instance, applies the physics of phase transitions to stock markets to explain financial crises (Jhun et al. 2018).

Computational – Sometimes the target can be so complicated as to be hardly tractable by any theoretical tool. Focusing on a simpler system that shares certain features in common could then allow one to gain insights into a complex system. For example, Schelling's (1969) "checkerboard" model uses a simple game to show how phenomena such as racial segregation in cities could occur as a result of a series of individual preferences – for example, a preference for living in a neighborhood in which one's race is well-represented. Here, the checkerboard functions as an analogue to gain insight about a social phenomenon (Sugden 2009; cf. Grune-Yanoff 2013).

1.2.2 An Antinomy of Scientific Reason

We can now reformulate the epistemological problem regarding the role of analogy in science anticipated in Section 1.1: In an evidentially impoverished context, under what conditions (if any) can there be justification in drawing an inference about the target from analogy? This question turns out to divide both philosophers and practitioners in stark ways. The fundamental aspect of this dispute concerns the legitimacy of analogical reasoning as a form of scientific reasoning. It is indeed a commonplace that analogies can lead us astray. Precisely because they are the easiest to reach, the analogy with familiar systems may set ourselves up to interpret the evidence in the way that we find most convenient, discounting important dissimilarities. In its attempt to reduce the unfamiliar to the familiar, then, the use of analogy may end up obscuring rather than illuminating a problem situation. This suggests that analogy should not be regarded as conferring justification. The predictable result of turning analogical reasoning into a scientific methodology would appear to be a picture of the world that is suspiciously "user-friendly" – misleading us into assuming uniformity when there is diversity. This seems to stand in contrast with the scientific attitude.

Following the above train of reasoning, what we will refer to as the *heuristic view* holds that nothing, except possibly experience, can serve as justification for analogical inferences in evidentially impoverished contexts (Agassi 1964, 1988; LaFollette and Shanks 1993; North 1981; Humphreys 2004; Crowther et al. 2019; Norton 2021; Azzouni 2022). Specifically, nothing except possibly the observed frequency of success in use can be an indication of the reliability of a source system for analogical inference. Frequency-independent criteria of plausibility for analogical inference cannot be formulated, or, even if they can, they do not track facts about when observations in one system can be a source of indirect information about another. Accordingly, the heuristic view sees analogy primarily as a ductile tool for scientific discovery, useful in creative tasks such as generating models and hypotheses, designing experiments, devising new problem-solving strategies, and so on, but without a proper justificatory role. Any appearance of a conclusion's receiving some justification from analogy is either illusory or owes to other facts in the background knowledge. The most radical version of the heuristic view embraces Popper's (1959) claim that hypotheses do not enjoy any kind of inductive justification whatsoever, including from analogy.

The rival doctrine is what we will refer to as the *inductive view* (Keynes 1921; Carnap 1980; Hesse 1966, 1974; Wylie 1985; Shelley 2002; Bartha 2009, 2022; Currie 2018; Nappo 2022; Holyoak 2025). Its defenders agree that analogies are often used as heuristic tools and that they are frequently in this function the source of false or misleading hypotheses, which occasionally put research astray. However, they insist that well-crafted analogies can also, sometimes, play a justificatory role in science. In particular, one function they sometimes have is providing a form of defeasible justification to hypotheses in evidentially impoverished contexts. The extent and nature of this justification varies: Some theorists favor expressing analogy's epistemic role in terms of *plausibility* (Bartha 2009) which is associated with the status of a hypothesis that meets a threshold of non-negligible credibility relative to its rivals; other theorists cash this out further in terms of *probability,* for instance in terms of an increase in probability associated with a hypothesis (Carnap 1980). What is common to all such versions is the claim that the consideration of an analogy can *rationally* and *non-negligibly* affect the credences or expectations that a scientist may form about the unknown. The most radical versions of the inductive view go so far as to regard analogy as a *criterion of confirmability* for scientific hypotheses: this is the view that Campbell (1920) defended specifically in relation to the role of analogy in theoretical physics (cf. Hesse 1966).

Wherever one's sympathies may lie, the methodological dispute is far from settled. Its echo is present across distinct scientific fields, with arguments on each side following common patterns (Wylie 1985). To counter skepticism, defenders of the inductive view typically appeal to facts about the scientific practice. A notable observation is that analogy frequently appears to be the source of significant expectations about hypotheses suggested by its means. This is not only true for practices such as extrapolating information about the toxicity of certain chemicals on humans from tests on mice, where there is a significant track record of past success, but also for cases in which no clear record exists. In physics, for instance, it has been noted that, when Coulomb presented the results of his "torsion" experiments on electromagnetism, many physicists already expected electrostatic charges to obey an inverse square law of distance due to the analogy with Newton's law of gravity. Similarly, the analogy between the masses of the Sun and atomic nuclei persuaded Bohr to seek the means to overcome the instability problem affecting his model. More generally, a consideration in favor of the inductive view is that entire research programs in science are sometimes initiated and pursued that are based solely on extensions of theories and methods that happened to have been empirically successful in some other context. This observation plays a central role in Kuhn's (1962) analysis of "normal science," which he understood as a process of puzzle-solving through extension of scientific exemplars (Bird 2008).

In addition to the expectations generated by striking analogies during research, a further aspect upon which defenders of the inductive view typically insist is that there often appears to be a normative structure accompanying analogical argumentation in science. This is clearest when examining live debates of working scientists. Shelley (2002), for instance, defends a version of the inductive view by documenting the aftermath of Louis Alvarez's (1980) assertion of an extraterrestrial cause to dinosaur extinction. Alvarez had compared the hypothetical asteroid crash hitting the Earth to the explosion of the volcano Krakatoa in Indonesia in 1883, whose debris had reached London. Alvarez's idea that dinosaur extinction could be caused by a massive asteroid crash with effects similar to the Krakatoa explosion was followed by a heated discussion in the scientific community over the plausibility of the Krakatoa-asteroid analogy. The similarities and dissimilarities between the effects were carefully analyzed, together with other evidence such as an unusual presence of iridium in the Cretaceous–Paleogene boundary of the terrestrial crust. As Shelley notes: "The scientists involved in the debate clearly thought that the counterarguments were relevant to the acceptability of [Alvarez's] hypothesis" (2002: 493). The importance given to assessing the analogy suggests that it was implicitly regarded as supporting the hypothesis.

Besides surface-level analysis, however, defending a version of the inductive view about analogy is a highly nontrivial task. Its defenders commit themselves to providing some account of when and how an analogy can play a justificatory role. Yet, even when we restrict our attention to analogical reasoning within a single discipline, this is by no means straightforward. Apart from the hard questions concerning justifying a form of inductive reasoning, an informative version of the inductive view requires deciding whether similarities between target and source should be allowed to be cross-type or restricted to within-type, as well as on the relevant approach to similarity aggregation. In each case, the inferences that are licensed by the theory change, and so does the story regarding the mechanisms behind their justificatory work. Moreover, to ground the conclusion of an analogical inference, one should expect the n properties over which the similarity basis is evaluated to have a relevant connection with the unknown property $n + 1$ to be inferred. However, judgments of relevance and irrelevance appear to vary with the background and purpose of the analogy, seemingly indicating that there is no fully general recipe for identifying strong analogical inferences that holds independently of the intended context of application.

In addition to the large variety of cases that a theory of analogy would need to consider, defenders of the heuristic view often point out how taking similarity as the driver of scientific inferences can often be misleading. Norton (2021: 124) illustrates this point through a historical example:

E2. Analogy between Earth and Moon

In *Sidereus Nuncius*, Galilei argued that the Moon has an irregular morphology similar to Earth. A key piece of evidence that he uses in support is the observation that: "On the fourth or fifth day after the new moon, the boundary that divides the dark part from the bright does not extend smoothly in an ellipse, as would happen in the case of a perfectly spherical body, but it is marked out in an irregular, uneven, and very wavy line. Now we have an appearance quite similar on the Earth at sunrise, when we behold the valleys, not yet flooded with light, but the mountains surrounding them on the opposite side to the Sun already ablaze with the splendor of his beams; and just as the shadows in the hollows of the Earth diminish in size as the sun rises higher, so also these spots on the Moon lose their blackness as the illuminated part grows larger." (Galilei 1610 [1957]: 53)

Enclosed in the quoted passage is a venerable example of an analogical argument: because the "wavy lines" observed on the Moon resemble the shadows that terrestrial mountains cast on valleys at dawn, it is plausible to suppose that the Moon has an irregular morphology like Earth's. But as soon as we look at the passage closely, Norton argues, the inadequacies of general theories of analogical inference become apparent. In his view, Galilei could have just as easily chosen the analogy with the shadow cast on the human face by the nose when light comes from a particular direction, without his conclusion's losing any plausibility. By Norton's lights, this suggests that the similarities with terrestrial shadows play a marginal role in Galilei's reasoning. If this is true, one must conclude that any theory that points to the similarities with terrestrial shadows as the main driver of the inference in Galilei's example likely misunderstands the latter.

The claim that the similarities of wavy lines with terrestrial shadows only play a marginal role in Galilei's argument can be denied. Regardless of whether one agrees with Norton on this specific example, however, the fact remains that an informative theory of analogical reasoning in science – one that, among other things, clarifies what is the role of the similarities in Galilei's reasoning and why terrestrial shadows are the "best analogue" for the lunar wavy lines – is difficult to come by. Conversely, defenders of the heuristic view have significantly more liberty in explaining away the seeming epistemic import of analogical argumentation in science compatibly with their conception. For instance, it may be argued that analogy is used in science merely to establish the *pursuit-worthiness* of hypotheses, where the latter is a status of deserving scientific attention that is detached from truth (cf. Nyrup 2020, who defends the more limited claim that sometimes analogy is used to establish pursuit-worthiness). In other cases, it has been noted that seeming uses of analogical inference are best reconstructed as inferences of some other (scientifically more respectable) type. For instance, some accounts of experimentation in the biomedical and social sciences deny that tests in laboratory settings can provide any inductive justification to hypotheses about non-laboratory settings (Plott 1991; cf. also Section 3.1.2). A different heuristic view, Norton's (2021) *material theory of induction*, rejects the need for explaining away putative instances of inductive use of analogy. On Norton's account, an analogical inference is justified just in case there is a "fact of analogy" (124) that grounds it (cf. also Section 2.2.4). Whether this obtains can only be established ex post, through empirical research, and not by any "logic" of analogy.

1.3 What's Next

Looking ahead, Sections 2 and 3 will address two key aspects of the disagreement between defenders of the inductive and the heuristic view about analogy in science. In Section 2, we will tackle the question of whether there are criteria that can be provided for when a conclusion reached by analogy is considered plausible in science. A significant part of the section will be devoted to presenting and evaluating the *structure-mapping theory* developed by Dedre Gentner and colleagues. The problem that we will raise for the theory is that its conditions are not sufficiently strict to tell apart the analogical inferences that are considered plausible in scientific research from those that are not. We will then assess alternative theories that include further constraints on plausibility. The options range from reviving Hesse's (1966) *two-dimensional theory* of analogical inference to more recent proposals, including Holyoak and Thagard's (1989) *multi-constraint theory*, Bartha's (2009) *articulation model*, and various *local approaches*. Our conclusion will highlight that all currently available theories suffer from important limitations.

In Section 3, we will take up the dispute between heuristic and inductive conceptions from a different angle. Setting aside the question of whether we can find general and informative criteria for plausibility by analogical inference, to what extent would such a theory help explain aspects of the scientific practice? As will be discussed, there are at least two contexts in which we may suppose a role for analogy at work: *model building* and *confirmation*. In the former, analogical inferences are employed in the way of articulating a model of a target. In the latter, evidence about an analogue system is sought to obtain indirect information about a target. It is not at all obvious that a general theory of analogical inference should find applicability in both contexts. In many of the examples that we will consider, the reconstruction whereby similarity constitutes the driver of the relevant inferences is only one of the available interpretations. Accordingly, our discussion will help clarify that the dispute between the inductive and heuristic views is not merely about the availability of a theory of analogy but about what constitutes proper scientific methodology.

Finally, Section 4 will discuss the thorny problem of justifying an inductive use of analogy, linking it to the more general problems of justifying induction. Our discussion will summarize the most promising attempts currently available and indicate directions for future philosophical research.

2 Criteria of Plausibility

A starting point for many versions of the inductive view is the consideration that scientific experts attribute different degrees of strength to analogies considered in contexts of research. Whereas many of them are perceived as a weak basis for scientific inference, others are taken in much higher regard, to the extent that entire research programs are sometimes built up and pursued on their basis. When a cogent analogy is proposed, the surrounding scientific community is often seen involved in an examination of its virtues and limits, in the comparison of its consequences with the known facts, and in the evaluation of its implications for research (Wylie 1985; Shelley 2002). But what makes a given analogy in scientific research cogent? Traditionally, philosophers and cognitive scientists have addressed this question by considering inferential behavior. That is, assuming that a cogent analogy generates more plausible analogical inferences, the question that they have attempted to answer by means of theory is what makes an *analogical inference* more or less plausible than another. Seeing that patterns of reasoning by analogy are often common in different sciences, these theories aim to establish criteria that apply across them.

As will emerge below, articulating a theory of plausible analogical inference in science that is both sufficiently general and precise is a highly complex task. Evidently, plausibility is partly a function of the modal profile of the conclusion: An analogical inference to the effect that a state of affairs is merely possible is more plausible, other things being equal, than one to the effect that the state of affairs actually obtains. However, the challenge for defenders of the inductive view is to provide criteria of plausibility when the modal profile of the conclusion is fixed. Existing approaches differ substantially in this regard. To facilitate the discussion, we provide here a short selection of existing answers, dividing them into *structure-mapping* (Section 2.1) and *multi-criteria* (Section 2.2) accounts. For each account in the selection, we will take note, first, of their level of *integration* with a working psychological theory of analogical reasoning and, second, of the *generality* and *precision* of the proposed criteria. We leave it to the reader to decide the weight of these evaluative dimensions.

2.1 The Structure-Mapping Theory

A highly influential account of analogical reasoning is Dedre Gentner's structure-mapping theory, as elaborated in several contributions (Gentner 1983, 1989, 2010; Gentner and Medina 1998; Gentner and Colhoun 2010; Gentner and Forbus 2011). The main hypothesis behind it is that the core of analogical reasoning lies in the dynamical recognition and processing of *relational matchings* between initially separate domains. The archetype of relational matchings is arithmetic proportions of the form "A:B = C:D," such as "1:2 = 3:6." Here, proportionality is given by a quantitative relation of preserving "the same ratio" between two pairs of numbers. Allegedly, the ability to recognize relational correspondences of increasingly higher order of complexity is at the basis of all analogical reasoning. Gentner (1983) uses the analogy between the atom and the Solar system as an illustration (see Section 1, E1). The representation of low-level commonalities like mass and motion in the analogy is said to favor the recognition of a higher-order matching in the relations between those properties, such as the match in the relation "greater than" holding between the masses in the source (Sun and planets) as well as between those in the target (nucleus and electrons). Further processing of the analogy may then lead to the recognition of relational matchings of an even higher order, such as the causal connection that relates the orbital motions of a planet around the Sun and electrons around the nucleus (Gentner 1989: 208; 2010: 753).

Gentner's theory is primarily an empirical theory about human reasoning. Agents performing analogical reasoning are viewed as involved in a (reiterative) process of generating mappings of relational structures and drawing inferences about missing correspondences. Specifically, a key aspect of Gentner's theory is that the initial stage of identifying the mappings "attends only to the *structure* of the knowledge representations for two analogues, rather than the content" (1983: 63). Earlier, in Section 1.1.2, we referred to this as a "structural approach" to similarity aggregation. Considerable evidence for this view derives from psychological experiments highlighting a distinction between *alignable* and *non-alignable* differences. Markman and Gentner (1993) demonstrate that, in testing conditions, people have greater ease in identifying differences between pairs of representations that are judged as semantically closer (such as "hotel" and "motel") than semantically more distant ones (such as "hotel" and "grapefruit"). This observation supports Gentner's theory insofar as one of its predictions is that, in performing alignments of relational structure, subjects would find it easier to identify alignable differences (viz., missing correspondences in a common structure of relations) over non-alignable ones (viz., any random difference). This is captured by

Gentner's maxim "*No difference without similarity*" (1993: 683), meaning that a cognitively prior step to recognizing differences is identifying structural mappings.

Gentner's structure-mapping theory is also a normative theory of analogical reasoning, in this sense: it implies that there are criteria for performing analogical reasoning that, if violated, result in poorly supported inferences. As such, the theory is also relevant to the epistemology of science. The ideal analogical performance is embodied in the "Structure Mapping Engine" (SME; Falkenhainer et al. 1989; Forbus, Gentner, and Law 1995; Gentner and Forbus 2011), a computational tool that, when given as input linguistic representations of two or more domains of entities with their associated sets of properties and mutual relations, first produces structural alignments among them and then formulates inferences based on pattern completion for the alignments that score highest by the evaluation criteria. To explicate the evaluation criteria, Gentner (2010) adopts the predicate calculus language. The elements of this calculus are names, standing for *objects,* and n-adic predicates, standing for *relations*. Monadic predicates represent *properties*, which are ascribed to objects in the domain of discourse: for example, MASS(x) attributes mass to object x. Non-monadic predicates represent relations between objects: for instance, ORBIT (x, y), which ranges over the two-place argument (x, y), stands for "x orbits around y."

The structure-mapping criteria embedded in the SME for selecting analogues are as follows:

(1) One-to-one correspondence: There is a bijection between the objects and relations instantiated in the source and in the target, meaning that for each object x or relation R in the source, there is one and only one corresponding object or relation in the target.
(2) Parallel connectivity: If a pair of n-adic predicates $R(x_1, \ldots, x_n)$ and $S(y_1, \ldots, y_n)$ are in a one-to-one correspondence, then the elements of their respective arguments (namely, $x_1, \ldots, x_n$ and $y_1, \ldots, y_n$, respectively) must also be in one-to-one correspondence.
(3) Systematicity: One-to-one correspondences connected by *constraining relations* (such as causal or explanatory relations) make for overall better structural matching.

Operationally, the SME takes structured sets of predicates representing two domains as input. It then repeatedly searches for ways to structurally align the source and target that satisfy the criteria above. Any mappings that satisfy criteria 1–2 are called *structurally consistent.* The third criterion, *Systematicity*, provides further ground for selecting among structurally consistent mappings by favoring those that carry the most explanatory or predictive power. In the language introduced in Section 1.1.2, SME features a structural approach to similarity aggregation until consistent matchings are found; after that, a relevance-driven criterion of systematicity is used to identify the best analogue. At the last step, the theory favors mappings that involve higher-order relations that can support explanation or causal inference (2010: 754). For example, having identified the relation GREATER MASS(x, y) that links nuclei to electrons, the SME may not yet have sufficient ground to select a Sun-planet model of the atom over (say) a parent–child model. But while various alignments may be identified that make the models equally structurally consistent, the relation ATTRACT(x, y), standing for "object x attracts object y," is a causal relation that is satisfied only by the pairs (nucleus, electron) and (Sun, planet). Because the relational matching tracks causal structure, the Sun-planet model is given preference on grounds of its greater "systematicity."

Several meritorious features of the structure mapping approach are worth noting, which make it a prominent competitor as an account of plausibility by analogical inference. First, the approach displays a high level of *integration* insofar as the implicit norms for analogical reasoning are couched within an empirically informed theory about the cognitive processes underlying human analogical reasoning. One consideration in its favor is the naturalness with which it explains the ease of accessibility of alignable differences compared to non-alignable ones. Second, conceived as a normative theory, Gentner's account possesses high *generality* and *precision*. The criteria provided are fully operationalizable and produce "best analogues" among multiple candidates whenever a representation of the target is given. Furthermore, the structure-mappings theory lends itself to a pluralistic account of the learning affordances of analogical reasoning in science. On Gentner's (2010) approach, systematic and structurally consistent mappings allow for different forms of learning: by *abstraction*, that is, by identifying common structures between distinct domains; by *contrast*, that is, by recognizing the differences between two systems with respect to corresponding properties; by *re-representation*, that is, by reconceiving the target or source in a different fashion (2010: 754); and, finally, by *inference projection*, that is, by imagining or supposing that the target displays further matchings with the source beyond those already observed.

However, some important problems remain. On a methodological level, there are inherent difficulties in testing the theory by its prediction about human subjects. Among other things, the theory's application to human performance is made difficult by the fact that human long-term memory tends to retrieve analogues based mainly on superficial similarity rather than structural consistency (Holyoak and Koh 1987). Gentner and Forbus (2011) acknowledge that their SME "does not currently model working memory limitations" (273) and is, therefore, unreliable in predicting which analogues human subjects will retrieve from memory. At the same time, when we consider cases in which retrieval of the source has already occurred and analogical mapping is involved, an obstacle to confirmation is that the linguistic representation of the source and target domain is hand-inputted in the SME to yield a preferred "best" analogue. One concern that this raises is that SME's predictions are not robust to choices of linguistic input, implying that the theory yields correct predictions only insofar as tailored linguistic inputs are used (Mitchell 1993).

On a more substantive level, there are problems with the structure-mapping account when understood as a normative theory of analogical inference. The main ones derive from the consideration that the theory seems hard to square with the judgments that trained scientists would issue over the plausibility of analogical inferences during research. Due to the prominence of the structure-mapping account in cognitive science, it is worth spelling out these issues in some detail, also considering the responses that its advocates have provided. The examples considered in what follows will serve to motivate the consideration of alternative approaches in Section 2.2.

2.1.1 Sufficient?

Structural consistency and systematicity do not appear jointly sufficient for plausible analogical inference. One set of purported counterexamples derives from what Hempel (1965) refers to as *nomic isomorphisms*, viz., correspondences in mathematical form between the laws of two entirely separate physical domains. In a highly mathematized field of science such as physics, it is often possible to show that very different quantities can, up to a certain extent, be viewed as formally interchangeable in distinct theories. Historically, some of these isomorphisms have played a fruitful role in the development of physics: earlier theories of heat, for instance, exploited formal analogies with the laws obeyed by fluids, and Joseph Fourier's work on heat conduction inspired George Ohm's articulation of the law of proportionality between electric current and voltage. The problem that these examples raise for the structure-mapping theory is that, even while scoring equally highly on the criteria of structural consistency and systematicity, not all instances of nomic isomorphism in science license equally plausible inferences by analogy.

A relevant observation is that it often matters in scientific practice whether a nomic isomorphism has emerged naturally or whether it has been intentionally sought and obtained via manipulation (Steiner 1998). An example from the domain of pure mathematics illustrates this difference.

E3. An algebraic argument for the Euler characteristic

Polya (1954: 43) notes that one can derive the formulation of the *Euler characteristics*, viz., the stable relation linking number of faces (F), edges (E), and vertices (V) of any regular polyhedron in solid geometry, by means of an artful manipulation of the analogous equation that holds in plane geometry. Starting from the relation holding for convex polygons, viz., V = E, one rewrites it as an alternate sum of respective numbers, as follows:

$$V - E + F = 1, \tag{1}$$

inducing the following equation for polyhedra, where "S" indicates the number of solids:

$$V - E + F - S = 1 \tag{2}$$

Despite having a true conclusion, the inference that (2) holds for any regular polyhedron based on the alternate sum similarity with (1) is weak (Cangiotti and Nappo 2023). Indeed, even though a structurally consistent and systematic matching is exhibited through the manipulation above, the algebraic similarity so exhibited is rather distant from the geometrical context of the problem.

We find another illustration of the general problem for structure-mapping in the following case:

E4. Formal analogies between electricity and heat

Figure 1 represents a section of electrically conductive material of length H and radius r, surrounded by insulation of thickness $r' - r$, such that the electric potential inside the cylinder is p and the one outside is p^*, where the arrows f indicate the direction of the electric field (Lange 2016: 351). It so happens that Figure 1 could equally well represent a cylindrical heat source of height H and radius r, surrounded by a layer of thermal insulation of thickness $r' - r$, such that the temperature inside the source is p and the temperature outside is p^*, and where the arrows f stand for the direction of heat flow. With appropriate manipulations of the relevant quantities, the equations that the two systems obey are formally the same.

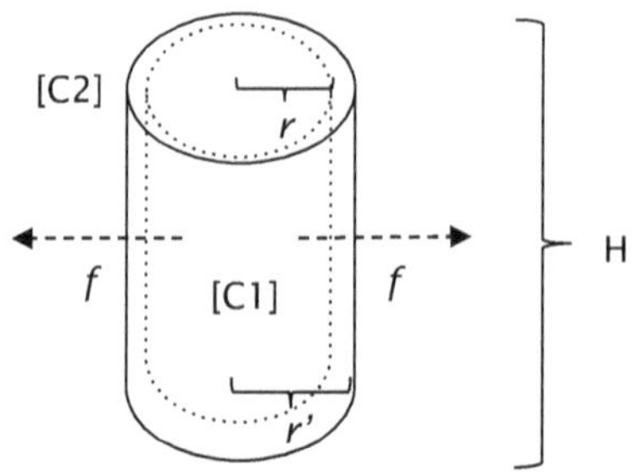

Figure 1 Analogy between heat and electricity

Despite the high structural consistency and systematicity of the analogy, the inference that disruptions to electric current flow would produce analogous effects in the thermal case is physically implausible. Again, this example suggests that a high degree of structural consistency and systematicity are sometimes insufficient to underwrite a plausible analogical inference.

In response, a defender of the structure-mapping theory might insist that inferences from equally systematic nomic isomorphisms are equally plausible despite not all finding equal assent by working scientists. However, this response marks an important separation between the descriptive and the normative level of assessment of analogical reasoning. The problem is not so much acknowledging that some nomic isomorphisms may generate "plausible" reasoning in some sense of the term. For instance, it has been argued that some formal isomorphisms may enjoy a weak form of *heuristic* plausibility within the field of particle physics (Fraser 2022). Rather, the issue lies with recognizing a form of plausibility of any kind to all inferences from structure-rich nomic isomorphisms. As Achinstein (1964: 328) has noted, if we deemed plausible any inferences from nomic isomorphism, we may just as well accept inferences about the color of swans based on their isomorphism with line segments: All we would need to do is to take the relation "has the same color as" to formally correspond to "is congruent with." What this suggests is that scientists are very unlikely to accept – whether implicitly or explicitly – that analogical inferences from equally systematic nomic isomorphisms are equally plausible. In this sense, there is a divergence between the deliverances of the structure-mapping theory and ordinary scientific judgment.

A different line of response on behalf of structure-mapping is suggested in Colhoun and Gentner (2009). They contend that plausibility should not be assessed with respect to individual analogical inferences – which is, in their view, far too variable across human subjects – but with regard to the totality of the analogical inferences that can be drawn from a given analogue. Hence, their considered claim is that greater structural consistency and systematicity induce analogical inferences that are *on average* more plausible, even though they are sometimes not so. In support of this claim, Colhoun and Gentner (2009) discuss empirical findings to the effect that, when measures of plausibility are global rather than individual, ordinary people's perceptions of plausibility are aligned with the prediction of the structure-mapping theory. Still, at least two problems emerge. First, the interpretation of the empirical evidence can be questioned: some other confounding factor may be present. Second, even conceding the explanation, there is a further question whether Colhoun and Gentner's result is robust when transferred to the scientific context: judgements of plausibility by scientific experts may differ from those of tested subjects.

2.1.2 Necessary?

Structural consistency and systematicity also appear unnecessary for the plausibility of an analogical inference. One argument is that superficial analogies are sometimes useful in scientific research (Bartha 2009; Currie 2013, 2018). In fields such as archaeology or medicine, for instance, establishing structurally consistent matchings is unlikely; often, one must resort to bare similarity. For example, when attempting to explain the function of an archaeological site tracing back to some ancient civilization, one can look for analogies with similar sites built by familiar civilizations, whose function is historically well-understood (Currie and Killin 2019; Nyrup 2021; Nappo et al. 2023). However, when the archaeological analogy is indirect, meaning that there is no historical or geographical connection between the target and source civilizations, one is often unable to identify a coherent and systematic structure of sociopolitical and religious relations in common. Still, sometimes bare similarity in features is used as evidence of similar function in archaeology.

A further argument is that sometimes an analogy is strengthened when elements of structural correspondence are shown not to hold (Lee and Holyoak 2008; Bartha 2009). In this case, the problem is that the structure-mapping theory judges systematicity only *after* identifying the most consistent matchings and without distinguishing contributing from preventative causes. For example, Darwin's *Origin of Species* elaborates a complex analogy between artificial selection exercised by breeders and natural selection. Although Darwin's use of analogy has produced an extensive discussion (cf. Gibson 2025 for a recent review), one observation shared by many historians and philosophers is that Darwin's conclusion – viz., that the high variety of traits observed in animal and plant species in nature is explained by natural selection – is made *more* plausible, and not less, by missing correspondences with artificial selection. In particular, there are differences in the scope of the selection process as well as in its duration: in breeding, selection occurs only with regard to certain traits and only on particular occasions, whereas in natural selection, environmental pressures are constant and more extensive; moreover, natural selection has acted for several millennia. These mismatches *strengthen* the analogical inference.

Again, the counterexamples can be rebutted. In response to the observation that superficial analogies can figure in plausible arguments in scientific research, defenders of the structure-mapping theory might note that those inferences tend to be weaker than inferences from highly structurally consistent matchings, due to the vagueness and elusive nature of bare similarity. However, this response implies that inferences from contrived formal isomorphisms, such as those considered in Section 2.1.1, tend to be more plausible than arguments from bare similarity. The inferential practice suggests that the opposite holds: contrived isomorphisms are rarely used by scientists in an evidential way, while bare similarity is frequently used evidentially (Currie 2018).

In response to the objection that requiring structural consistency and systematicity is unnecessary for plausibility, defenders of the structure-mapping account may explain counterexamples away by appealing to their pluralistic account of learning from analogy. In Darwin's example, for instance, they might note that a common structure between artificial and natural selection is a prerequisite for learning "by contrast," that is, for drawing information from the differences between artificial and natural selection. On this view, it is the existence of an analogous mechanism for selection that explains the variety of individuals in both cases – and only then the differences in scope and length with artificial selection may enable one to conclude that the high variety of individuals observed in nature is compatible with natural selection. However, this seems to push the problem back a step, as it is not obvious at one point one should judge that a structural matching ought to be used in the "inference" mode rather than in "learning by contrast" mode.

2.2 Adding Constraints

The emerging problems for the structure-mapping theory may suggest looking elsewhere for criteria of plausibility. Perhaps the theory is under-constrained. Or, perhaps, an entirely different starting point is needed. There are various proposals in the literature on where to look for an alternative, each with its advantages and drawbacks. Here we briefly discuss four accounts.

2.2.1 The Two-Dimensional Theory

Defenders of the inductive view who look for a more demanding conception of plausible analogical inference than the structure-mapping account may be drawn to seminal work from Hesse (1966) (Chen 2020; Nappo 2022, 2025; Cangiotti and Nappo 2023; Holyoak 2025). The central claim in Hesse's *two-dimensional* theory is the idea that the evaluation of an analogical argument in science should occur simultaneously alongside two dimensions. First, there is the *horizontal* dimension, which concerns the evaluation of the similarities attributed to the source and target. For instance, in the analogy between an atom and the Solar system, one evaluates the extent to which nuclei are like the Sun and electrons are like planets, respectively. Here, key aspects of the evaluation are the questions: *How* similar are the source and target? And are the respects in which scientifically respectable? Secondly, there is the *vertical* dimension of the analogy, which concerns the evaluation of the relation that the known similarities bear with the merely predictive similarities. For instance, considering whether electrons may orbit around the nucleus like planets around the Sun leads to asking: Is it likely that similar causal relations hold between nucleus and electrons?

Starting from an analysis of historical examples of scientific disagreement, Hesse (1966) goes on to extrapolate the implicit norms that guide plausible analogical inference in science alongside the horizontal and the vertical dimensions. An important condition that she defends on plausible analogical inference in science is a precursor of Gentner's *Systematicity* criterion:

(1) Causality: There must be reason to think that the properties $p_1, \ldots, p_n$, that a target shares with the source are caused by, or are conducive to, the property p_{n+1} that figures in the argument's conclusion.

For instance, there is no plausible inference from the observed bodily similarities $p_1, \ldots, p_n$ between a person and her identical twin to a conclusion p_{n+1} – say, that they both have flat feet – unless those similarities somehow make p_{n+1} more likely. Hesse (1966) put this in terms of the requirement that there must be "causal relations" among the salient properties (68), adding that "causal" should be intended "in some scientifically acceptable sense of the term" (68). However, she never meant that the presence of the causal relation between $p_1, \ldots, p_n$ and p_{n+1} must be *known* (cf. Bartha 2009: 38). Her condition only requires that there is reason to suppose a causal relation in the target of the same kind as that which holds between the source's properties.

More relevant to our present discussion is Hesse's claim concerning the horizontal dimension of evaluation of an analogical inference in science:

(2) Materiality: The similarities figuring in the premises of an analogical argument must be "material" or "pre-theoretic" (1966:67).

Although this requirement has been interpreted in wildly different ways (Dardashti et al. 2019; Norton 2021), an example that Hesse draws from the political realm (1966: 63) clarifies her core idea. Inferring that citizens owe respect and obedience to the State in the same way as children do toward their fathers is a bad analogical inference, on Hesse's view, because "there does not seem to be any horizontal relation of similarity between the terms [viz., children and citizen, on one hand, and father and State, on the other] except in virtue of the fact that the two pairs are related by the same vertical relation" (1966: 63), such as the relation "provider-for" (63) that links father to child. In other words, Hesse's point is that the political analogy is implausible because the similarities that the argument invokes (between father and State and children and citizen, respectively) do not exist *before* and *independently* of the structural correspondences that the argument itself makes salient. A similar condition on plausibility holds, on Hesse's view, in the scientific context: the similarities that drive an inductively strong argument must be recognized as genuine respects of similarity *before* and *independently* of the analogy. Hesse (1966) initially suggests that this is equivalent to requiring that the horizontal relations in an analogical inference must be "observables" (93); however, her later discussion (93–96) indicates that her main concern is telling apart analogical inference that stand on an independent recognition of similarities and those that *generate* the similarities by imposing a specific way of looking at the target system.

When read in accordance with the above, friendly interpretation, Hesse's condition on horizontal similarity is a serious candidate for diagnosing what may be missing from the structure-mapping theory. Its elegance lies in providing a single diagnosis to two pathologies related to analogical inference. One is the practice of *gerrymandering predicates* to inflate a weak inference, as when one notes that a sapphire observed after 2,100 is likely to have the same chemical composition as an emerald observed today on grounds of both being "grue" (Goodman 1955). Insofar as "grue" does not belong to the language of the scientific community when the analogical argument is given, the proposed similarity fails the materiality test and cannot be used in inference (Hesse 1974). A related case is given by inferences from *pure nomic isomorphisms*, such as those that we have seen as making trouble for the structure-mapping theory (see Section 2.1.1, E3). As Hesse (1966) notes, the discovery that there is a similarity in form between the equations describing the domain of elliptic membranes and those which describe the balancing act of an acrobat is "useless for prediction" (69), insofar as no antecedent similarity exists between the two *independently* of their respective quantities' serving as placeholders for some arbitrary mathematical structure.

Casting Hesse's condition in terms of a notion of materiality *relative* to the accepted language of a scientific community also serves to mitigate a concern that one may have with the condition: viz., the fact that, by imposing that the similarities are recognized before and independently of the structure of relations holding between the properties, it appears to give "our naïve judgments of resemblance a privileged status" (Bartha 2009: 43), in a way that seems problematic. On a friendly interpretation of Hesse's condition of materiality, however, what counts as a "pre-theoretic" similarity depends on the state of knowledge and common assumptions of the scientific community *at a given time* (Nappo 2025). When confronted with the objection that there is no independent ground from which to draw the distinction between "pre-theoretic" and "theoretic" properties, Hesse (1966: 15) readily admits this ("it can plausibly be argued that there is *no* descriptive statement, not even the "blue-here-now" beloved of sense-data theorists"), adding that the distinction cannot be but a "pragmatic one, relative to the assumption of a given language community" (15). She then goes on to mention an example of analogical inference in quantum mechanics, where "it need not trouble us that what stands in the place of the observation language here is . . . the language of classical physics . . . for we have already agreed that what counts as an observation language is pragmatically relative" (48). This suggests that, for Hesse, the distinction between "pre-theoretic" and "theoretic" respects of similarity is not set in stone but changes with the growth of scientific knowledge. In a mature science, pre-theoretic respects of similarity may be far from "naïve," having been trained by experience and consolidated theoretical developments.

The sensitivity of the evaluation to background assumptions also plays a role in the final clause in Hesse's theory, which pertains to both the horizontal and vertical dimensions of evaluation:

(3) No Essential Difference. There must be no known differences in the "essential" properties and causal relations of source and target (89).

On Hesse's view, what counts as an "essential difference" cannot be given an a priori definition but requires the crucial input of context. Hesse (1966: 89) gives the example of early physical theories of heat. Even though modeling heat as a fluid could capture many aspects of its behavior, by the mid nineteenth century it was understood that heat is not a conserved quantity. At that point, to continue using the fluid model for prediction involved a decision about whether conservation was an essential property. That decision, on Hesse's view, rests "partly on judgments of what is causally essential to the model, and partly on the current availability of alternative models" (Hesse 1966: 92). In effect, the fluid model was abandoned partly in concurrence with the development of thermodynamics. Together the requirement of *Materiality, No Essential Difference* prevents classifying as plausible inferences from contrived or "forced" formal analogies (see Section 2.1.1, E3).

While Hesse's two-dimensional account continues to be a source of inspiration, several limitations of the theory should also be recognized. First, the theory is based mostly on conceptual clarification and rational reconstruction of instances of scientific reasoning. In this sense, it features *low integration* with an empirically grounded cognitive theory of analogical reasoning. As it stands, Hesse's theory also suffers with regard to *generality* and *precision*. First, it addresses a single mode of learning by analogy, viz., inferring a property of the target from a property of the source ("inference projection"). Second, her approach to similarity aggregation is thoroughly "relevance-driven" and does not countenance more precise rules of inferential conduct. Most notably, Hesse's theory does not provide guidance with regard to how to reason with *multiple analogues*, that is, distinct candidate sources for a given target system. It is not possible for the theory to explain, for instance, that terrestrial shadows cast by mountains were a better analogue than the shadow cast by the nose on the face to support Galileo's argument for the Moon's irregular morphology without appealing to context (cf. Norton 2021: 124). Accordingly, while the theory can distinguish between plausible and implausible *sets* of candidates for analogical inference, further selection within the sets is to be accomplished without theory guidance.

Evidently, turning Hesse's insights into a complete theory of analogical reasoning in science will require further work. Meanwhile, defenders of structure-mapping may consider incorporating a *Materiality* constraint into their theory to better address the problem of nomic isomorphisms.

2.2.2 The Multi-Constraint Theory

A different approach is defended in Holyoak and Thagard (1989) and further articulated in several works (Holyoak and Thagard 1995; Lee and Holyoak 2008; Holyoak 2025). Their *multi-constraint theory* retains many aspects of the structure-mapping theory, most notably the hypothesis that one of the hallmarks of human cognition is the ability to spontaneously identify relational mappings between initially separate domains. At the same time, the multi-constraint theory marks some notable departures from Gentner's theory, effectively accommodating several insights from Hesse's work. The core idea behind the multi-constraint theory is that in the process of making and assessing analogies, human reasoners implicitly aim to strike a balance (which the authors refer to as "coherence") between three dimensions. The first dimension is a purely *structural* one, whereby analogies are identified based on their capacity to approximate an isomorphism (or one-to-one correspondence), as in Gentner's approach. The second dimension is what they often refer to as the *semantic* dimension of similarity, whereby – much like Hesse's requirement of *Materiality* – the strength of an analogy is evaluated partly based on the prior similarity that is perceived between source and target. The third dimension is a *pragmatic* one, whereby the analogy is evaluated more positively if it incorporates elements that are important to its intended purpose.

The idea of finding a balance between distinct dimensions finds a concrete realization in a computational model called "Analogical Constraint Mapping Engine" (ACME). The conditions that ACME incorporates to identify best mappings and draw inferences from them include Gentner's (1989) criteria of *One-to-one correspondence* and *Connectivity*, which we summarize here as:

(1) Structural consistency: There is a structurally consistent mapping between source and target (in Gentner's 1989 sense).

Additionally, Holyoak and Thagard impose the following further constraints:

(2) Semantic similarity: The mapping maximizes "intuitive" (or "prior") judgments of similarity across the properties put in correspondence.
(3) Relevance: The mapping incorporates elements (such as causal relations) that are viewed as important to achieve the reasoner's inferential goals.

The semantic and pragmatic dimensions introduced by the theory provide additional flexibility in explaining a variety of psychological observations, including how human reasoners often place differential emphasis on structural correspondence or semantic similarity depending on purpose and context (Richland et al. 2006). Moreover, the new constraints help respond to the objections faced by structure-mapping. First, the appeal to the dimension of semantic similarity and to the reasoner's purpose may help account for the different perceptions of plausibility for analogical inferences based on nomic isomorphisms (Holyoak and Thagard 1989: 301). The intention of using a system that obeys mathematically analogous laws to draw inferences about an unfamiliar target may raise the standards for semantic similarity between the systems, which could lead a scientist to reject a nomic isomorphism as implausible despite its high structural consistency. Second, the appeal to the reasoner's purpose may also help explain why sometimes an element of mismatch between two systems (as in Darwin's disanalogies between natural and artificial selection) can improve an analogical inference. This can be so if some of the properties that are attributed to the source (e.g., artificial selections' depending on the time and labor of a human breeder with finite resources) are preventative causes with regard to the phenomenon to which one is putting attention (e.g., the impressive variety of traits found among domesticated animals).

Compared to rivals, the multi-constraint theory scores highly on *integration* and *generality*: It seeks to not only describe the process of making analogies as a complete reasoning process but also its underlying psychological mechanisms (and even hinting at its neural basis; cf. Holyoak 2025). However, critics have noted that its generality comes at a cost in terms of *precision*. Understood as a normative theory, the claim that balance across multiple constraints equates to greater plausibility is vague; in many cases, it fails to yield any meaningful prediction (Bartha 2009). As Holyoak and Thagard (1995) admit, there is no single recipe for weighing against one another different constraints on coherence. Methodologically, this leads to the worry that the theory overfits the data, only managing to recover the judgments that human reasoners provide after the fact. This can still be a cause of significant dissatisfaction for those who seek more precise (and falsifiable) predictions and concrete normative verdicts from a theory of plausible inference.

In recent work, Holyoak (2025) distinguishes the general level of providing a theory of analogical reasoning – which includes describing the various uses that reasoners may make of analogies – and the specific case of giving an account of the evidential use of analogy in the scientific context. Interestingly, when considering the latter use, Holyoak defends Hesse's conditions of *Causality*, *Materiality*, and *No Essential Difference*. One interpretative question that remains open from Holyoak's (2025) recent proposal is whether he views Hesse's criteria as merely an instance of the multi-constraint theory, which can be derived once the purpose of the analogy is set to be an inductive (or evidential) one, or whether some auxiliary premises about scientific methodology must be introduced to derive Hesse's normative criteria. Nevertheless, Holyoak's suggestion remains interesting insofar as it suggests a possible convergence between two independently motivated theories of analogical reasoning – one based in psychology, the other in epistemology.

2.2.3 The Articulation Model

A different starting point to the problem of plausibility is given by Bartha's (2009) *articulation model*. Like Hesse, Bartha adopts the philosophical methodology of rational reconstruction rather than the empirical tools of cognitive science. But while Hesse's focus is the level of articulating reasoning in relation to scientific communities and their practices, Bartha's focus is the dialectical context of argumentation that sees a *proponent* or *advocate* opposed to a *critic* in the evaluation of an analogy. The aim of the theory is to make explicit the norms that this practice of argumentation is subject to when the proponent and critic are jointly invested in settling an empirical question. The idea is that the articulation of these norms may be viewed as detached from a description of the cognitive mechanisms supporting analogical reasoning. As the label "articulation model" suggests, Bartha's guiding idea is that making a conclusion plausible involves the proponent's taking up a burden to *clarify* their analogy (cf. Polya 1954: 13). This involves making clear which properties of source and target are going to be relevant to the conclusion of the analogical argument – what Bartha calls the analogy's "*prior association*." Having proceeded in the clarification task, the plausibility of the conclusion of an analogical argument is then assessed mainly on the dimension of "*potential for generalization*" and, in particular, whether there are *critical disanalogies* that impede the extension of the source's properties onto the target.

Compared to the accounts previously reviewed, which emphasized the need for simultaneous satisfaction of independent constraints on plausibility, Bartha's account offers a relatively straightforward *linear* model of evaluation, which depends crucially on the ability to distinguish the sets of properties that are relevant to the conclusion of the argument from irrelevant ones. For instance, in proposing an analogy between atoms and the solar system, the proponent's role consists in first clarifying the features of the source that are relevant to explaining the stability of the solar system: presumably, features such as bodies' masses and motion in relation to one another. This first, stage-setting phase of clarification of the "prior association" in the analogy's source is then followed by the identification of the relevant features of the target that match those of the source, relative to the analogical argument's conclusion, and in the assessment of the importance of the features of the target that are not shared by the source. In our example, this could involve evaluating to what extent the differences in mass that one finds between the Sun and planets approximately match those in the target, as well as evaluating whether the instability of electronic behavior predicted by classical electrodynamics represents a critical difference.

Another feature of Bartha's account is that it aims to provide normative criteria of plausibility governing specific types of analogical arguments in science. Hence, the theory includes various levels of descriptions. At the most general level, an analogical argument passes the threshold of plausibility on Bartha's model when the following conditions are sequentially met:

(1) Prior Association: There is a clear connection, in the source, between a set of properties $p_1^*, \ldots, p_n^*$ that the source is known to share with the target and one or more properties p_{n+1}^* that the source also possesses.
(2) Overlap: The set of known similarities with the target (among the $p_1, \ldots, p_n$ identified as relevant by the prior association) is non-empty.
(3) Potential for Generalization: There is no critical disanalogy, that is, some dissimilarity which undermines the possibility of extending the source's prior association to the target.

Bartha (2009) goes on to specify these criteria further depending on the *type* of prior association and the *direction* of inference. A type of analogical argument that Bartha countenances is that of the purely "correlative" kind, where the prior associations – what Hesse would have called the "vertical relations" – are not causal but purely statistical. Contra Hesse, Bartha argues that correlative analogies can sometimes yield plausible reasoning (cf. Nappo 2020). With regard to the analogies in which the prior association is causal or explanatory, Bartha further proposes to distinguish between two *directions* of analogical inference. "Predictive" analogical arguments infer similar effects from similar causes. An example is arguing that a cure will be safe and effective on humans because it has been tested on mice, which are similar in biological features. "Abductive" analogical arguments, conversely, infer similar causes from similar effects. An example may be reasoning that a common gene is present in both mice and humans, because of their similar response to a drug (cf. Castellani 2024). A virtue of Bartha's theory is that it provides separate indicators for evaluating abductive and predictive arguments, resulting in significant precision.

A somewhat more vulnerable aspect of Bartha's account concerns his *Overlap* condition. The liberality of this requirement is evident when compared to Hesse's *Materiality*. Whereas Hesse imposes that some similarities between source and target must be recognized before and independently of any analogical inference, Bartha claims that the evaluation of similarity should only occur *after* the prior association (the vertical relations) has been clarified. This claim exposes his articulation model to the same objection that we have raised for Gentner's theory, stemming from analogical inferences from nomic isomorphisms. One weakness of these inferences, as we said, plausibly lies in the fact that the similarities emerge only after refashioning the target in a specific mathematical form – an operation that is easy to accomplish for far too many targets. After considering the problem, Bartha notes that a return to Hesse's *Materiality* would be a mistake, since "arguments based on formal analogy have . . . been extremely influential in physics" (2009: 43). Instead, he proposes to integrate *Overlap* with a further proviso:

(2*) Physical Significance: The similarities that drive the analogical argument must have "known physical significance" (2009: 221).

However, some issues with 2* remain. First, the requirement is less general than *Materiality*, as it exclusively applies to analogies in physics. This leaves out inferences from nomic isomorphism in disciplines other than physics. Second, the decision to evaluate the physical significance of the similarities only after the clarification of the prior association may still raise problems. A template for counterexamples may be furnished by returning to the algebraic argument for the Euler characteristics (see Section 2.1.1, E2). The algebraic similarities that figure in the argument, such as instantiating an alternating sum pattern, clearly have "mathematical significance." Yet, a key weakness of the argument is that those similarities play out in an extraneous geometrical context. Analogous problems may affect the condition of Physical Significance when properly scrutinized.

As a philosophical theory, Bartha's model is characterized by *low integration* with a cognitive theory of analogical reasoning. At the same time, the *generality and precision* of the theory are high. In particular, the articulation model extends to cases in which there are multiple candidate analogues for a target. Bartha (2009) defends a condition of *Independence*, stating that the plausibility of a conclusion achieved by an "optimal" (2009: 142) analogical argument is independent from other potential candidate analogues; "suboptimal" arguments, instead, do not provide plausibility. The distinction is achieved by means of a ranking, as follows: optimal arguments (such as Galileo's comparing the Moon's wavy lines to terrestrial shadows) are those in which the set of relevant similarities is the largest among the candidates and which feature no critical disanalogies; suboptimal arguments (such as comparing the wavy lines to the nose's shadows) fail to satisfy either of these conditions. It follows that, in case of multiple optimal arguments, each supports their conclusion independently of the others; conversely, when an argument is found to be suboptimal relative to others, it fails to support its conclusion. Although Bartha (2009) recognizes that this is merely an outline of an account, it is worth highlighting this feature compared to rivals. As we have seen, Hesse's (1966) theory lacks a proposal about multiple analogies, whereas both structure-mapping and multi-constraint theories provide them.

2.2.4 Local Approaches

An emerging set of accounts aims to make progress in the debate by considering how proprietary criteria for plausible analogical reasoning emerge from the study of specific applications. It is important to clarify in what sense these approaches represent a novelty in the debate. An idea that practically all general theories of analogical reasoning share is that discipline-specific information can play a role in determining plausibility or implausibility for analogical inference. This is most evident in the case of the requirements that we have referred to (depending on the specific theory) as *Systematicity*, *Causality*, or *Prior Association*. It is a feature of these constraints that they are anchored to an understanding of what constitutes scientifically acceptable causal or explanatory relations. This may well be both time- and discipline-dependent. For instance, when discussing the scientifically admissible senses of "causal relation," Hesse countenances a "form of attack [to analogical inference] based on a disagreement about whether the connections asserted in the model do after all deserve to be called causal" (81). As Hesse's (1966: 84–85) reconstruction of disputes in pre-Darwinian biology suggests, divergences over plausibility sometimes come down to whether the connections between the properties of the analogy's source have a claim to be causal, relative to the state of knowledge in a given scientific field.

At the same time, defenders of local approaches do not deny that, at some level of abstraction, we may account for the plausibility of analogical inferences by the observance of a few vague but general criteria (Wylie 1985; Currie 2018). They may also recognize that those vague criteria become more informative when interpreted in accordance with the understanding internal to a given discipline – as when we specify what an acceptable "causal relation" is through field-specific information. Their main contentions can be discerned into a methodological and a substantive component. Methodologically, defenders of local approaches argue that a historical and philosophical perspective should try, first and foremost, to uncover the diversity of the practices that surround the use of analogy in science. On a more substantive level, these theorists claim that the insights one can gain from such a bottom-up perspective reveal tensions and, sometimes, inconsistencies with general theories of analogical reasoning, which make it likely that achieving a sufficiently informative theory of plausibility for the scientific domain will have to be articulated in a more disunified and piecemeal fashion that purely top-down perspectives allow. In this sense, we interpret local approaches as aiming to steer a middle ground in between defenders of general theories of analogy, on the one hand, and, on the other, theorists (such as Norton 2021) who deny that any ex ante criteria for plausibility of analogical inference in science can be provided.

By way of illustration, let us provide two noteworthy examples of such local approaches. The first example comes from Currie's (2018) work on the historical sciences. Taking up the problem that historical remains may often be unique pieces with no analogue elsewhere in other remains (Wylie 1985), Currie argues that paleobiologists often successfully circumvent this problem by means of creative patchworks of several imperfect analogues (cf. Wylie 2011). Currie refers to this strategy as *exquisite corpse*, borrowing the term from a surrealist parlor game. One of his examples is:

> ***E5. Exquisite corpse***
> The saber-toothed metatherian *Thylacosmilus atrox* – a rather unique species which lived through much of the Cenozoic in South America – is compared to *Smilidon fatalis* in its bite mechanisms, in that both favor the use of neck over jaw strength to bite preys, as well as to current-day bears in killing style, in that both aim to first secure the prey by immobilizing it. By mentally recombining the features of two distinct and imperfect analogues, paleobiologists can arrive at a plausible picture of *T. atrox*'s actual bite and killing style.

There are tensions – perhaps even inconsistencies – between the way paleobiologists adopt the exquisite corpse strategy and general theories of analogical reasoning. For one, some general theories (such as Hesse's 1966) do not countenance criteria for evaluating multiple analogues. Moreover, general theories that do incorporate provisions for multiple analogies may fail to account for the specific inferential practice that Currie (2018) describes. For instance, the exquisite corpse strategy can be viewed as a counterexample to Bartha's (2009) abovementioned condition of *Independence* for multiple analogies. It appears that sometimes two analogues may rank differently in terms of relevant similarities and yet be used *simultaneously* to provide plausibility to a hypothesis about a target – contrary to what the *Independence* condition holds.

Another – rather different – example of a local approach emerges from Nersessian's (2008) work on conceptual innovation in science. Whereas Currie's project is mostly concerned with extrapolating field-specific patterns of plausible analogical inference that escape the sight of general theories, Nersessian's focus is on the cognitive dimensions of reasoning with and through analogues. In her view, even the theories of plausible analogical inference that are most highly integrated with cognitive science – such as Gentner's structure-mapping theory and Holyoak and Thagard's multi-constraint theory – do not provide sufficiently realistic descriptions of the cognitive processes that underlie the selection of an analogue. This claim is exemplified through several case studies of conceptual change (cf. also Nersessian and Chandrasekharan 2009), including:

> ***E6. A fluid analogy to electricity and magnetism***
> To provide a mathematical vest to Faraday's ideas about electricity and magnetism, J. Clerk Maxwell's paper "On Faraday's Lines of Force" (1856) proposes an analogy with fluid dynamics. On this model, the *direction* of an electric or magnetic force is represented by a line of force. For instance, a line of force is said to pass between the north and south pole of a magnet. In order to represent the *intensity* of the force, we are asked to conceive of lines of force, not as geometrical lines, but rather as "fine tubes of variable section carrying an incompressible fluid" (157). As it is well-known in the case of fluids, the velocity of the fluid in a tube varies inversely as the section of the tube. Maxwell's idea, then, is to represent the intensity of a given electric or magnetic force at any given time by the section of the tube carrying the fluid, or (which amounts to the same thing) by the velocity of the fluid along the direction of its line of force. Through this analogy, Maxwell shows that many of the macroscopic properties of both electricity and magnetism can be accounted for.

According to Nersessian (2008), general theories of analogical reasoning underplay the aspect of *construction* of the analogue as a key element in the scientific process. In Maxwell's case, the fluid that functions as a model for electromagnetism is not an ordinary fluid but rather some kind of *intermediary* – a useful idealization of the behavior of fluids that he utilizes for the specific purpose of modeling electromagnetic phenomena. As Maxwell confirms: "the substance here treated of must not be assumed to possess any of the properties of ordinary fluids except those of freedom of motion and resistance to compression (Maxwell 1890: 160; cf. Nappo 2021; Suarez 2024). Underlying Nersessian's argument is the concern that general theories of analogical reasoning are too narrowly focused on norms of plausible reasoning and on separating plausible from implausible inferences, while missing out on the more creative components of the process of constructing new conceptualizations as observed in the actual thinking and work of scientists. Similar remarks apply to uses of analogy in engineering (cf. also Knuuttila and Loettgers 2014b).

Predictably, the strategy of sharpening the evaluation criteria by going local also has theoretical costs. Insofar as local approaches exploit acquaintance with specific fields or patterns of thinking, they inevitably trade in *generality* for greater *precision*. This can lead to two sorts of worries. First, there is the question of how "local" we ought to go: when is a description of the inferential practice sufficiently precise? In addition to this, there is a concern of fragility: Specific patterns of plausible reasoning are likely more vulnerable to internal contestation than patterns that apply in whole disciplines and across several fields of science. Consequently, it is important for defenders of local approaches to specify the *scale* of their concern and to justify their choice. With regard to the dimension of integration with a psychological theory, the two examples of local approaches provided above stand at rather opposite sides of the spectrum. Currie's discussion focuses on specific norms and patterns of plausible analogical inference that are relevant to a given scientific discipline or subfield, while remaining neutral on the possibility of integration with a psychological theory of analogy. By contrast, Nersessian aims to explicitly anchor her framework in the cognitive sciences, treating scientific reasoning as an extension of ordinary mental processes.

2.3 Discussion

The literature aiming to uncover the criteria of plausibility of analogical reasoning shows a large variety in starting points and methodologies. Our survey has covered only a portion of the spectrum of alternatives, but other approaches would be worth considering as well (Weisberg 2013; Walton et al. 2012; Macagno et al. 2017). Despite some notable convergences among extant proposals, the prospects for identifying a theory of plausible analogical reasoning remain unclear. The structure-mapping theory satisfies various putative desiderata, including high integration with a cognitive science account of human analogical reasoning, high generality, and high precision. However, it often seems hard to square with the judgments regarding plausibility that emerge from various scientific contexts. This leaves us with a hard choice, between acknowledging that judgments of plausibility emerging from the scientific practice are misled and, instead, attempting to recover those judgments by adding further constraints. The former route is unappealing for theorists who aim to defend an epistemic role of analogy in science, since it matters to their philosophical project scientific judgments of plausibility are guided by a coherent form of logic. Conversely, as previously discussed, taking the latter route is riddled with difficulties.

The state of uncertainty permits both pessimistic and optimistic outlooks. On the more pessimistic side, Norton (2021) has argued that we look in vain for an account that settles the problem of plausibility for analogical reasoning in science. For him, the problems shown by extant theories are too significant to expect that they will find a resolution. On a more optimistic outlook, the successes and convergences among extant accounts are important, indicating that further research is not in vain and that many breakthroughs are still possible (Genta 2020; Bartha 2022).

3 Analogy at Work

Let us return to the divide between the heuristic and the inductive view of analogy in science. As we discussed, defenders of the inductive view commit themselves to providing an account of when an analogy can possess a broadly epistemic role, in justifying, to some extent, empirical hypotheses. This has led to the investigation of the previous section. As we have seen, the prospects for identifying a satisfactory account that serves the purposes of the inductive view are still unclear. However, even assuming that a satisfactory account of plausible analogical inference can be identified, the challenges for the inductive view do not end here. It is part of this view to also show that an account of plausible analogical inference can *explain* features of actual scientific methodology. If the proposed account does not contribute to understanding how scientists use models in scientific investigation in an interesting variety of cases (beyond some toy examples) and why the inferences that they draw from them may be justified, the claim that analogy plays a role in the production of empirical knowledge becomes unsubstantial and rather marginal to scientific methodology. This is what we may call *the problem of extendibility* for the inductive view. Again, the fundamental asymmetry with the heuristic view must be stressed. Insofar as on the latter view analogy only plays a role in activating creative thinking and exploring new avenues, there is no pressure on its defenders to reconstruct various aspects of the scientific practice as cases of how analogy is "at work." The burden is solely on defenders of the inductive view.

The problem of extendibility is not addressed properly in the recent literature in philosophy of science. Defenders of theories of plausible analogical reasoning tend to view analogy at work wherever science uses surrogate systems or makes comparisons. However, as we will see, there are often open questions whether and how the theory latches onto the features of the specific inferential practices in science. Although it is not required that the theory apply to all apparent instances, we should expect it to be informative in at least an interesting subset of them. A source of additional difficulty is that there are at least two broad contexts in which analogy may be "at work" in contexts of scientific research, each possessing distinctive features. First, analogical reasoning can play a role in the construction of models of real-world targets, either by directly helping to shape the model or by helping identify an already existing source. Here, we will use "*model building by analogy*" to refer to this activity. Once a model is built, analogical reasoning can play a role in transferring the properties of the source to the target. Typically, observations about the model's properties are undertaken, which carry over, via similarity, to support hypotheses about the target. We refer to this activity as "*confirmation by analogy.*" (Elsewhere, in Nappo 2022, the distinction between model building and confirmation has been drawn in terms of a *non-incremental* versus an *incremental* notion of confirmation by analogy).

Of course, due to the variety of inferential practices observed in science, discussions of the problem of extendibility are necessarily limited in scope. For instance, the full structure of a model built by analogy varies with its domain of application and its purposes: models in highly mathematized sciences tend to have a richer formal structure, while models in experimental sciences have a richer empirical content. Moreover, the level of access to the target system differs significantly, ranging from contexts in which there is no empirical access whatsoever to the target, to less evidentially impoverished contexts in which some empirical evidence, either direct or indirect, may be available. In what follows, we have selected four examples that, for their variety, are representative of how the problem of extendibility may arise when reconstructing the actual inferential practice in science. (For another relevant discussion, see Castellani 2024 on convergence arguments). Specifically, the next two subsections investigate purported cases of analogy at work in confirmation. The main question that will concern us is whether, and to what extent, an analogy may contribute to the justification of the relevant inferences. The last two subsections, instead, focus on cases of model building. There again, we will ask whether analogy is at work in the building of the model and whether, additionally, analogy can be considered partly responsible for the expectation that the model makes successful predictions about the target.

3.1 Confirmation by Analogy

From a logical viewpoint, model building precedes confirmation by analogy: in order for hypotheses about the target to be confirmed by similarity with a given model, a model is needed in the first place. For the sake of our discussion, however, we begin with cases of confirmation by analogy. Let's start from a case that illustrates the problem of extendibility in a neat way.

3.1.1 Analog Simulations

One of the most exciting developments of contemporary cosmology is the hypothesis, originally advanced by Bekenstein (1972), that black holes may exhibit thermodynamic behavior. Work by Hawking (1976) has given large impetus to what had initially appeared as mere speculations. Using a semiclassical model, viz., a model that approximates quantum mechanical behavior within a setting that obeys the principles of general relativity, Hawking derived the prediction that black holes emit a form of radiation analogous to heat. By Hawking's calculations, a black hole could be ascribed a "temperature" proportional to its surface gravity. Unfortunately, direct empirical evidence for the thermal nature of black holes is not forthcoming. Background cosmic radiation is far too strong to let us detect the Hawking effect – unless we could find some extremely small black hole. At the same time, a "trans-Planckian problem" arises with Hawking's prediction: looking carefully at the calculations, it appears that emitted photons should come out of ultrahigh energy regimes in their distant past, with wavelengths well below the so-called Planck scale. The assumptions of general relativity that are incorporated in Hawking's model are expected to break down at those ultrahigh energy regimes, casting doubt on the reliability of the model's predictions.

Based on work by Unruh (1981), an active field of experimental physics utilizes *analogue simulations* to indirectly test the existence of Hawking radiation. Here is a simplified summary:

E7. Analogue simulation of a black hole

The core idea is that accelerated fluid flow can produce a sonic black hole (or "dumb hole"), viz., a region from which sound waves cannot escape, mirroring the behavior of light in a gravitational black hole. The geometry instantiated by the fluid system is mathematically analogous to the space-time curvature geometry described by general relativity. By a Hawking-style derivation from the physical description of this system, a prediction of

> Hawking-like sonic radiation can be derived (Unruh 1981). Crucially, the fluid system would exhibit an analogue of the trans-Planckian problem: the classical fluid dynamics used to predict sonic radiation are assumed to hold invariably for all energy scales, even though the atomic theory of matter is known to prevent this. In virtue of this observation, Unruh and colleagues argued that verifying sonic radiation in an analogue simulation would provide an important hint that black holes emit Hawking radiation. Currently, a thriving research field in experimental physics aims to test sonic radiation in analogue black holes (Barcelo 2019).

Would the verification of sonic radiation provide evidence of Hawking radiation? Prima facie, such an inference would be based on analogy: A source system, on which we can readily perform experiments, is identified as an analogue of the target, based on alleged similarities between them (cf. Section 1.1; Sterrett 2017). However, Dardashti et al. (2017) challenge this reconstruction. They argue that the evidential weight of these simulations does not rest on analogical reasoning. Instead, they propose that the evidential weight of black hole simulations depends on the availability of *universality arguments*: viz., independent evidence that the thermal behavior emerges from a set of physical conditions that are realized by vastly different physical systems. Because evidence can be provided to the effect that the black holes and their analogues belong to the same "universality class" (2017: 76), their physical differences become irrelevant. Therefore, Dardashti et al. claim that sonic radiation can confirm the prediction of Hawking radiation.

Dardashti et al.'s universality-based account exemplifies an attempt at resisting the claim that a specific methodology adopted by scientists rests on mere analogy. However, as it happens in other cases in which an alternative to the analogical interpretation is proposed, the methodology behind Dardashti et al.'s universality-based account has also been challenged. An objection against it is Crowther et al.'s (2019) observation that, because as scientists cannot access black holes, they cannot verify whether they satisfy the conditions for belonging to the same universality class as their fluid analogues. Consequently, the claim that universality lies behind confirmation from analogue simulations does not appear to be well-grounded. Nappo and Cangiotti (2025) add that, if the belonging to the universality class is merely postulated, any observation of analogue realization would only provide *enumerative* evidence for Hawking's prediction: it would be like inferring that the next ball drawn from an urn is white because the previous two have been white. If correct, these criticisms would rescue reconstructions in terms of analogy (Bartha 2022).

Assuming that the analogical interpretation is correct, further questions emerge. First, how is the similarity basis established, given that empirical access to black holes is precluded? Second, does the alleged similarity basis justify placing additional confidence in Hawking's prediction if sonic radiation is observed? Crowther et al. (2019) and Nappo and Cangiotti (2025) disagree on the answer to these further questions. Crowther et al. view the analogy between black holes and their fluid analogues as merely formal; hence, they claim that observations performed on accessible fluid systems are incapable of supporting the prediction of Hawking radiation. By contrast, Nappo and Cangiotti defend the thesis of inductive support by arguing that the research program of analogue simulations is driven mainly by an expectation, supported by physical argument, that black holes and their fluid surrogates present physically relevant similarities.

The ongoing debate over the epistemology of black hole simulations is useful to illustrate what we have called the problem of extendibility. Even though the case may strike as an instance of the use of an analogue system to gain indirect information, the claim that similarity is the basis of the relevant inferences can be denied – and sometimes plausibly so, especially if empirical access to the target system is impossible or if an alternative reconstruction grounds the inferential practice on more secure epistemological foundations. In such a case, we can say that the initial reconstruction in terms of analogy has been "*overruled*." The task of the defender of the inductive view then consists in either (a) showing that the relevant inferences are reducible to some known pattern of plausible analogical inference, thereby vindicating the face value analogical interpretation or (b) admit the case study as a type of inference that the theory does not cover, either because it should be treated separately or because it does not yield plausible conclusions. Clearly, advocates of the inductive view should use option (b) with parsimony, on pain of compromising the argument that analogy plays a non-marginal role in the production of scientific knowledge. However, there are circumstances in which the best option for inductive theorists may simply be to deny that a theory of plausible analogical reasoning should cover the example.

3.1.2 Extrapolation

The previous section has discussed a case in which it is not obvious *whether* a general theory of analogical reasoning should apply. But the problem of extendibility also concerns *how* the theory latches onto the particular uses of analogy. We illustrate this through cases of *extrapolation*, or *analogue experimentation*, where empirical access to the target is limited but still possible. As a premise, we note that some accounts of extrapolation reject the idea that tests in an accessible source can provide inductive evidence for hypotheses about some other domains, focusing instead solely on their capacity to refute incorrect general theories (Plott 1991). For all its value in providing methodological insights about possible forms of experimental design, this view remains less common, as it struggles to provide coherent reconstructions of the logic behind many realistic examples of extrapolation practiced in science (Guala 2005). A prime example would be:

> ***E8. Extrapolation from animal models***
> Protocols for assessing the effects of substances on the human body – ranging from medical drugs to toxins – typically require a multi-stage empirical process. This process relies heavily on analogical reasoning to bridge the gap between experimental conditions and human biology. In many cases, the target is not involved and hypotheses about human patients are tested indirectly using animal models. For example, to extrapolate the claim that toxic aflatoxin B1 causes liver cancer, biologists may study its behavior in mice. After observing that aflatoxin B1 induces DNA damage and cause liver cancer, they may look for correlations between aflatoxin exposure and liver cancer in humans.

Although it is widely accepted that mice-to-human extrapolation rests on analogical reasoning, important methodological issues remain. In particular, it is not clear how one should establish and reason from the similarities between the target and animal model, in view of the fact that humans and animals also feature substantial physiological and genetical differences. Can a general theory of plausible analogical inference provide any sound and informative methodological guidance?

Let us immediately note one way in which the general theory may be useful. In their argument against animal experimentation in the biomedical sciences, Lafollette and Shanks (1993) assert that the practice rests on *Causal Analog Models*. For such models to produce reliable inferences, they further argue, no causally relevant disanalogies should exist with their targets. This condition is rarely met in animal experimentation. While some quantitative differences can be corrected based on the different physiological (e.g., body weight and surface size) and biogenetical properties (e.g., DNA composition), in many other cases even similar properties (including homologue proteins, which are used to claim common evolutionary origins for humans and primates) can serve different functions in each species or be fulfilled by different causal mechanisms. Lafollette and Shanks (1996) conclude that scientists ordinarily lack evidence to ground the scientific validity of their animal experimentation. Here, a common defense of the experimental practice points out that the bar of evidence has been raised too high: while it cannot provide certainty, animal testing can increase confidence that medical treatments will be effective in humans. A general theory of analogical inference can then be invoked to argue that causally relevant disanalogies do not automatically defeat inductive support by analogy (Steel 2008).

But while the theory may be useful to counter skepticism, several other methodological questions remain unsettled. One of them is particularly important to experimenters: exactly which similarities and dissimilarities should we check to ensure that an extrapolation is sufficiently justified? Here, two types of approaches are possible. On a *qualitative* approach, assessing an extrapolation ultimately comes down to evaluating the strength of an analogical inference from model to target; guidance on the relevant aspects for extrapolation is therefore based on the consideration of observed similarities and background knowledge. Guala (2005, 2010), who defends this approach, explicitly refers to Holyoak and Thagard's (1995) multi-constraint theory of analogy as inspiration for his account. An argument for the qualitative approach is its applicability to many different cases of extrapolation. In addition to analogue experimentation on animal models, Guala (2005) also considers problems of extrapolating results obtained in laboratory settings to a more general population. He illustrates this with an example from experimental economics:

> ***E9. From laboratory auctions to real-world auctions***
> In the 1970's, evidence from oil leases in the Gulf of Mexico indicated that, in auctions based on a first-price, sealed-bid leases would tend to be overpriced, leading to the hypothesis that auction winners are "cursed" to be the bidders that most highly over-estimate the lease's value. Kagel and Levin (1986) tested the winner's curse hypothesis in the laboratory. In the experiment, auction rules are the same as in real-life contexts; however, laboratory conditions can be controlled in that the value of the lease is known and further information can be given to the bidders to revise their bids.

As Guala stresses, extrapolations to real-world auctions are possible inasmuch as one observes that, in both contexts, similar changes in the amount and type of information provided to bidders are correlated with similar changes in auction price. Here, the choice of which similarities and dissimilarities to focus on for the extrapolation ultimately depends on one's contextual understanding of the underlying causal mechanisms. For Guala, this also means the extrapolation can only provide "subjective" (2010: 1075) support, contingent upon one's state of knowledge.

If the methodological guidance offered by the qualitative approach is perceived as insufficient or overly vague, *quantitative* approaches provide a more regimented framework. Some relevant examples include Pearl and Bareinboim's (2014) *do-calculus* approach and Van Eersel et al.'s (2019) *latent class* approach. Below, we focus on Steel's (2008, 2010) *comparative process tracing* approach, which has emerged as the most prominent and scrutinized account of extrapolation in the recent philosophical literature (cf. Parker 2010). Comparative process tracing seeks to anchor extrapolation in the identification of similarity and dissimilarity alongside distinctive markers, or "fingerprints." Steel (2010) defines these fingerprints as specific downstream effects that can only result from a particular set of upstream causal properties within the similarity basis. Much like a telltale symptom can reveal a pathology, these fingerprints serve as an indicator of a shared causal mechanism. For the confirmation of "upstream" stages of a causal process, it is thus sufficient to check key "downstream" stages upon which any upstream difference between model and target must leave its mark. Steel offers a formal treatment of this approach via causal graphs representation, which involves building a model of the source system in which all relevant causal variables are represented, alongside relations of dependence from other variables in the model (Steel 2010). Under this representation, justifying extrapolation becomes a matter of checking whether the causal model for the source system matches (where all "fingerprints" are properly represented) the causal model for the target system.

Steel's approach proves particularly robust in the context of biomedical research, where it offers a compelling response to the skepticism of Lafollette and Shanks (1993) regarding causal analog models. In the study of Aflatoxin B1, for instance, the difficulty of mapping the entire metabolic pathway from exposure to liver cancer is circumvented by focusing on a specific fingerprint: the presence of DNA adducts in the blood. Because these adducts are specific by-products of the carcinogenic mechanism, observed agreement between animal models and human subjects serves as a key indicator that the underlying causal process is identical across species. Steel (2010) argues that comparative process tracing extends to the case of extrapolating results from a laboratory setting, revisiting Guala's analysis of outer continental shelf auctions. Through his formal treatment, Steel demonstrates that the "winner's curse" hypothesis can be supported by systematically ruling out alternative causal explanations for the observed bidding behavior.

The shift toward formalization invites the question whether such models truly improve upon qualitative approaches or merely mask the same underlying dependence from analogical reasoning. As Guala (2010) observes, even the most sophisticated formalization of extrapolation requires an antecedent choice regarding which causal variables are relevant in the tested model – a choice that remains contingent on a researcher's background knowledge and contextual understanding. While our discussion does not aim to adjudicate the dispute between these rival approaches, the tension between them is relevant to assessing the problem of extendibility. Guala's qualitative approach is derived from a general theory of analogical reasoning – the multi-constraint theory – and aims at broad application. Conversely, comparative process tracing represents an attempt to define a logic of analogy for a more limited set of inferential practices, which include extrapolations in biomedicine and assessments of external validity in the social sciences. Although comparative process tracing still relies to a significant extent on similarity, this approach seeks to move beyond the qualitative recommendations of general theories of plausible analogical reasoning, providing instead a more formally regimented procedure of assessment.

3.2 Model Building by Analogy

So far, we have discussed cases in which a model has been identified, and analogical reasoning is used to underwrite the passage from observations about a model to hypotheses about a target. However, as anticipated, sometimes analogy is invoked in the construction of entire models and not just in the formulation or confirmation of individual hypotheses about a target. Model building by analogy can take different forms. One approach is to build the model directly, based on its similarity with the target. Here, one-to-one similarities are supposed to be within-type. This approach is linked to the similarity conception of scientific representation, which has been systematically discussed in the literature (Frigg and Nguyen 2020). Alternatively, one can build a model by analogy in an indirect way. For instance, one can first identify another source as a surrogate and then take a successful model of such a system as a basis for a model of the target. In this case, known as *model transfer*, the similarities are typically cross-type. Yet another indirect case of model building by analogy presupposes that the models incorporate similarity metrics. *Clusterization* is a geometrical method to subdivide the dataset into clusters, wherein elements of the cluster are close to each other on average. In the next two subsections, we deal with indirect forms of model building – by far the most interesting for assessing the problem of extendibility.

3.2.1 Model Transfer

When theoretical limitations reflect a lack of knowledge on how to model a target system, a common practice among scientists is the attempt to transfer concepts and methods from theories successfully developed for a given application to the target – a mode of investigation sometimes referred to as "*model transfer*" (Knuuttila and Loettgers 2014a, 2019). The underlying presumption appears to be that, since a model has worked in one context, the same model should also work for another context possessing similar features. As the transfer can occur across rather distinct domains, for example, from the physics of critical phase transitions to stock markets, similarities between the relevant properties are typically cross-type. The questions we are interested in here are the following: What exactly is the mechanism for model transfer? And to what extent does it actually rest on analogy? These questions clearly intersect the central issue of this section, concerning the extent of the applicability of general theories of plausible analogical inference in science.

To inform our discussion, let us take an illustrative example, much discussed in the literature:

> ***E10. Transfer of the Ising model***
> The Ising model was originally built to describe the behavior of magnets in statistical physics. At the microscopic level, magnets are represented by a large number of spins placed in a lattice, with two possible states depending on whether they point in the up or in the down direction. Each spin interacts with its neighbors through a magnetic potential. A paramagnetic phase corresponds to the spins being oriented randomly. If the temperature decreases below a critical point, all spins suddenly get oriented in the same direction, leading the system to transition into a phase called "ferromagnetic." A "critical phase transition" such as this is macroscopically described in terms of a quantity called magnetic susceptibility. By the Ising model, this quantity follows a power law and it diverges at the critical point when a phase transition occurs. Interestingly, the same model is used as a basis for representing coordinated collective behavior in several other scientific fields, from socioeconomics to biology and neurosciences. For example, the JLS model for stock market crashes by Johansen, Ledois and Sornette (2000) is inspired by the behavior of critical phase transitions. Here, traders in a market play the role of spins in a magnet: they interact with their neighbors and can be in either one of two states, depending on whether they sell or buy stocks. Market crashes occur when all traders end up selling stocks at the same time, just like all spins tend to point in the same direction for a ferromagnet. To account for financial data, the model derives a macroscopic equation for the evolution of price, which depends on a quantity called hazard rate. This quantity follows a similar power law as magnetic susceptibility.

From an epistemological viewpoint, relevant questions concern the justification for transferring elements of the Ising model to build the JLS model – and whether the same justification holds for other cases of model transfer. This is the subject of a flourishing debate. Knuuttila and Loettgers (2019) and Herfeld (2024) distinguish *analogy-based* and *template-based* approaches. In the former, similarities between the target and source drive the transfer of the model of the source system to the target (Bokulich 2015; Zuchowski 2019). Here, the common structure of the models is viewed as reflective of the similarity basis between the target and its analogue. Accordingly, on this approach, the justification to build a model by transfer depends on the extent and relevance of the similarity basis. For instance, in the JLS model there are high-level formal similarities between macroscopic quantities, such as magnetic susceptibility and hazard rate, but there is also a lower-level analogy between the microscopic components of magnets and financial markets, in that both spins and traders have two available states (up\down and sell\buy), and they also share a similar causal mechanism for collective coordination due to their local interactions. General criteria of plausible analogical reasoning in science can then be used to evaluate to what extent we can expect market stocks to behave like magnets under temperature changes.

Championed by Humphreys (2002, 2004, 2019), the template-based approach provides a different perspective on the practice of model transfer. On Humphreys' specific version of the approach, the justification for model transfer can only be given by checking that the construction assumptions of the new model are "representationally correct," which requires empirical knowledge of the target. For this, Humphreys argues, analogical reasoning is of little justificatory weight. The practice of model transfer should rather be given a proprietary reconstruction, in terms of the identification and exchange of *templates* of theories. In Humphrey's language, templates are general representational devices that enable the construction of scientific models. The reason for their wide application has to be located primarily in their high generality, rather than in some mysterious uniformity of natural and social phenomena. Model transfer can accordingly be decomposed into two main steps. First, a formal template is extracted from an existing theoretical model by abstraction. Second, it is applied to the target's context via a reinterpretation of theoretical terms. The process of abstraction in the first step is what provides templates with generality, in the sense of making them independent from any specific theory. Two other features of templates aid the second step: Their flexibility makes them amenable to applications in different contexts; their tractability facilitates the construction of models. On Humphreys' view, producing models in this way has the sole purpose of generating hypotheses that can be tested. At least in principle, then, scientific model transfer can be entirely automatized (Humphreys 2004).

Starting from Humphreys' original contribution, different versions of the template-based approach have emerged, possessing different implications with regard to the problem of extendibility. Herfeld (2024, 2025) interprets the approach as sharply separated from the analogical one, arguing that generality rather than similarity is key to template-based model transfer. Her examples of such transfers include the application of game-theoretic tools across the sciences, whose wide use owes in large part to the generality of the mathematical tools (together with a series of contingent historical factors). By contrast, drawing on examples such as the transfer of the Ising model, Knuuttila and Loettgers (2014a, 2014b, 2016) propose a different account. In their view, the template-based approach should not reduce the wide reach of some scientific models (such as the Ising model) solely to computational and contingent aspects – for example, the generality and familiarity of the mathematical methods employed. Rather, one key reason for speaking of "templates" rather than "analogies" in conceptualizing model transfer is its drawing attention to *how* some highly tractable models are reutilized across fields. Specifically, an analysis of various scientific episodes illustrates the tentative and dynamical nature of the process of reapplying known models in new domains; in such a process, the source domain often works more like a than as a *foil* than as an indirect representation of the target. That is, learning can be activated not only by the similarities, but rather (and often primarily) by the recognition of the *dissimilarities* between the "template" and the new context (Knuuttila and Loettgers 2014b).

At the same time, an important implication of Knuuttila and Loettgers' view is that template-based reasoning cannot be entirely divorced from analogical reasoning, either. Commenting on Humphreys's (2004) account, Knuuttila and Loettgers (2016) first note that in some cases (like the Hopfield model for neural networks) one cannot really check whether models are empirically correct, and so relying on their similarity with empirically correct models of the source is useful for the purpose of justification. Second, Knuuttila and Loettgers argue that the need for analogical reasoning at some stage of the process of model transfer is evident when a model is transferred across very different contexts, for example, from physics to neuroscience. In these cases, one should not conceive of model transfer merely in terms of an abstract formal structure but also, and most crucially, in terms of the transfer of a common way to conceptualize phenomena. This leads to their definition of "model template" as a formal template *equipped with* a conceptual framework (Knuuttila and Loettgers 2016). Such a framework is informed by empirical observations on the distinct systems, as required to provide "initial plausibility" (2016: 398) to the transfer. For example, the Hopfield model of auto-associative memory in cognitive neuroscience is built by transferring the template and conceptualization of the Sherrington–Kirkpatrick model of spin glasses in physics, which is, in turn, built via the template of the Ising model. This suggests that the process of transferring templates across scientific fields must still be informed by analogy at some level.

Knuuttila and Loettgers' hybrid account raises the question of exactly how analogy-based evaluation of model transfer interacts with a template-based one. In many cases, considering similarities and dissimilarities seems necessary to distinguish more and less successful transfers. To underscore this point, Palacios and Jhun (2023), along the lines of Jhun et al. (2018), show that the JLS model for financial crashes has important disanalogies with the Ising model. While a higher-level similarity between the models is grounded on formal isomorphisms between macroscopic properties such as magnetic susceptibility and hazard rate, the lower-level similarity between the microscopic constitutions of magnets and stock markets is less structured. In effect, whereas the Ising model gives a mathematical description of how spins mutually interact through a magnetic potential, the JLS model does not incorporate details of how traders interact with each other. Technically, the former introduces a well-defined Hamiltonian for the local interactions between spins, whereas the latter only posits that traders interact with their neighbors but does not specify a Hamiltonian to model the tendency of traders to imitate each other when buying and selling stock. As Palacios and Jhun contend, this disanalogy is responsible for the relative lack of predictive success of the JLS model compared to the Ising model. In fact, without considering how traders interact, the model loses the ability to make accurate quantitative predictions.

At the same time, another example from so-called "econophysics" suggests that there is also a sense in which dissimilarities are necessary for a transferred model to have predictive success in a new context. Models of quantum finance borrow the formal treatment of uncertainty that is successfully applied in quantum theory to deal with the uncertainty surrounding financial markets, for example, concerning the evolution of the price of certain assets. Here, the analogy is more abstract than in the case of the JLS model, in the sense that such models do not mainly rest upon low-level similarities between microscopic components of quantum systems and financial markets. Rather, the guiding idea is to establish high-level, formal analogies between relevant quantities. In particular, Baaquie (2004) derived an uncertainty relation between the price of an asset and its rate of change in time, which is akin to the one famously derived by Heisenberg for position and momentum of a quantum system: in the former, the commutator of such quantities is equal to Planck's constant multiplied by an imaginary unit, while in the latter it is equal to the square of price volatility, which is a real number. However, as Arioli and Valente (2021) have pointed out, there is a crucial difference between these uncertainty relations in that, contrary to the quantum case, in finance there does not appear to be any imaginary unit. That is what renders models of finance applicable to stock markets, for they can thus make empirical predictions about (say) the evolution of prices without having to include peculiar quantum-like phenomena whose description requires the imaginary unit (e.g., interference patterns of electron beams in the double-slit experiment). In this context, dissimilarities prove as important as similarities in the process of model building.

The question posed above, concerning the interaction of analogy-based and template-based evaluations in the operation of model transfer, relates to other current discussions in the literature. For instance, several works in the philosophy of interdisciplinarity address whether and how we should try to evaluate operations of model transfer across scientific disciplines. A recurrent criticism in bridging disciplines is that a model sometimes appears to be *forced* on the target, rather than naturally applying to it. Considering model transfers in the field of economics, Mäki and Marchionni (2011) identify a form of "model imperialism," whereby the domain of application of successful models in one field is expanded to incorporate the domain of application of the models in another "colonized" field, including its underlying ontology. Bradley and Thebault (2019) contrast this approach with "model migration," whereby elements of the source are exported to the target, but the underlying ontologies remain distinct. An example comes from Bayesian phylogeographic models in historical linguistics, in which methods for tracing the origin of virus outbreaks through geographic and molecular sequences of data are reapplied to model how languages spread in the course of history. While model imperialism implies that the source ought to share the same properties with (the model of) the target, model migration allows for the target and source to remain separated, and thus similarities between their properties are cross-type.

Again, the problem of how interdisciplinary model transfer should be evaluated can be viewed as a further illustration of the problem of extendibility. Defenders of a purely template-based approach would reject the claim that a general theory of analogical reasoning should adjudicate on which instances of model transfer are to be deemed successful in science. By contrast, defenders of an analogy-based approach (or even a hybrid approach) would insist that analogical reasoning should play a role in evaluating instances of model transfer across domains, recognizing the need to scrap or readjust the models in light of critical dissimilarities between the systems.

3.2.2 Similarity Metrics and Clusterization

Another way to build a model by analogy rests on a different approach to similarity compared to the practices reviewed so far. In data-intensive fields such as genomics and climate science, the sheer volume of information often precludes the use of traditional causal mapping. In these contexts, scientists may appeal to *similarity metrics* to identify analogues within a given dataset, thereby adopting what in Section 1.1.2 we dubbed the "geometrical approach" to the aggregation of similarities. These metrics are informed by topological distance functions that tell us how close a source is to a target with regard to a set of relevant properties. An overall score is then computed based on a chosen similarity measure. In this way, even massive datasets can be processed using algorithmic methods. While this methodology has received little attention in the philosophy of science literature, it has a growing number of applications in current research, ranging from the clusterization of patients in biomedicine and the search for spatial and temporal analogues in climate science to the methodology of analogical proportions used in machine learning. Of course, to survey all the different applications of similarity metrics in computational science goes far beyond the scope of this subsection. We limit ourselves to two examples, both of which raise the question whether geometrical analogue methods can be considered an autonomous methodology for scientific inference, independent from a general theory of plausible analogical reasoning.

The first example comes from climate science, specifically from the domain of adaptation:

> ***E11: Spatial climate analogues***
> The methodology of "spatial climate analogues" purports to identify an existing analogue of the future climate of a target city by means of computer simulations run under various emissions scenarios (Hallegatte et al. 2007). To achieve this, a small set of climate indicators – typically, temperature means, number of cooling or heating days and amount of precipitations – and aggregated via similarity metrics. The objective is to translate matches in statistical patterns into tangible reference points for climate adaptation. By mapping the projected climate variables of a target city – such as Paris in the year 2080 – researchers can scan current global meteorological data to identify a "best-fit" geographical match. For instance, a simulation might reveal that the future climate of Paris most closely resembles the current climate of a Mediterranean city like Cordoba. Urban planners and policymakers can therefore observe existing infrastructure, agricultural practices, and public health strategies in the analogue city as a source of information for their future requirements.

Valente et al. (2024) point out a series of critical issues with the methodology of spatial climate analogues. First, a *problem of multiple metrics*: There are multiple inequivalent similarity metrics that can be adopted. Grenier et al. (2013) review six distinct metrics, showing that they do not agree with each other in their respective predictions. What is more, the outcomes of simulations depend on the climate models one adopts. For instance, the analogue of the projected climate of Paris is Bordeaux under the ARPEGE model but Cordoba under the HadRM3 H (Hallegatte et al. 2007). Second, there is a *problem of noncausal correlations:* since the geometrical comparison of the target and source reveals only statistical patterns across data, there is no guarantee that they track any causal process related to urban climate. Third, there arises a *problem of difference*, in that "average" similarity tends to wash out individual similarities and dissimilarities. Indeed, even a high similarity score does not preclude that there may be large one-to-one dissimilarities with regard to some relevant indicators. For instance, the city of Paris is rather different from putative spatial analogues like Bordeaux and Cordoba in terms of size, urban planning, and surrounding landscape, and so the adequacy of these analogues can be disputed. For these reasons, Valente et al. (2024) recommend addressing these shortcomings through appeal to local knowledge, so that the deliverances of similarity metrics can be integrated with qualitative expert assessments.

Interestingly, other applications of similarity metrics present common epistemological challenges. For instance, Boniolo et al. (2023a) discuss a computational technique called *clusterization* that is widely applied in various fields of science, especially precision medicine, for the classification of individuals in a population into similarity classes. In this case, similarity is typically computed over an extremely large number of known properties of the patient population. An example is:

E12. Clusterization of patients for prediction of treatment

A population of cancer patients may be clusterized to predict the most appropriate medical treatment for individual patients. A large dataset of omic features (genome, transcriptome, epigenome, proteome, etc.) is built spanning the molecular profiling of all patients. Then, a technique of clusterization known as k-Nearest Neighbors is typically applied. It consists of two steps. In the *learning step*, the patient population is partitioned into *k* clusters, by associating individuals with their nearest "neighbors" in terms of a formal similarity metric. The dataset is divided into a training set (80% of the sample) and a validation set (20%). The ability of the computational model to classify the cohort of patients into the correct tumor types is then evaluated based on statistical criteria, namely accuracy, precision and recall. In the *predictive step*, predictions about prognosis, course of the pathology and medical treatment are formulated based on the behavior of the other patients in the cluster. For instance, patience may be assigned a probability to develop a specific cancer type based on the frequency of patients in the cluster that developed the same illness.

Again, clusterization is affected by many of the same general problems indicated above. First, outcomes are not robust across metrics. Based on a toy model, Boniolo et al. (2023a) show that the size and composition of clusters depend on the distance function one adopts to find the "nearest neighbor." Second, they demonstrate that formal similarity metrics merely track statistical correlations between patients, and without additional constraints one cannot know whether the observed patterns are driven by any similar causal mechanism. Finally, it may as well happen that the target patient has non-negligible differences with respect to some omics with the individuals in the assigned cluster, while being more similar to individuals in another cluster. Therefore, the typical proviso of general theories of plausible analogical reasoning, whereby no "critical" disanalogies should exist with the target, may be violated in geometrical applications.

The problems noted above for spatial climate analogues and clusterization raise the question as to the reliability of the inferences one can draw from formal similarity metrics. For our purposes, it is worth highlighting how the examples provide further instances of the problem of extendibility. In practice, it is open to debate if we should reconstruct the above applications of geometrical approaches as instances of an autonomous scientific methodology with predictive use or if a general theory of analogical reasoning should be invoked to explain when and why the use of similarity metrics may hold evidential value. What is certain is that establishing the geometrical approach as an autonomous methodology with predictive force necessitates further conceptual and empirical development. While computational techniques for the search for causal patterns may, in principle, help address the problem of noncausal correlation (Pietsch 2021), other limitations currently appear more resistant to solutions. The problem of multiple metrics, for instance, requires the formulation of criteria to justify the selection of a specific similarity metric (or a class thereof) over its alternatives. To deal with the problem of difference, researchers could establish margins of acceptability, or tolerance thresholds, regarding the discrepancies allowed within a candidate analogue. By imposing such limits, even a source system that achieves a high aggregate similarity score would be disqualified if it possesses specific divergences that, according to ordinary analogical reasoning, would be regarded as critical disanalogies.

4 Justification

Let's take stock. As we discussed in the previous sections, the debate among the inductive view about analogy and the rival doctrine, namely the heuristic view, unfolds on different levels and across different scales. In Section 2, we have considered the problem of articulating conditions regarding when scientists assign plausibility to analogical inferences. This task is by no means easy; one may add that the varied and ubiquitous nature of analogy in science is such that restricting attention to a specific scientific discipline hardly makes the problem significantly more manageable. In Section 3, we have further observed that working instances of inductive use of analogy can also be interpreted as inferences of some other kind, leaving the burden on defenders of the inductive view to show that their account of plausibility does explanatory work. In each case, the inductive view is far from being fully vindicated. This leaves us at an apparent standstill. Of course, *if there is an inductive use of analogy*, the next question to ask is what is that grounds it: in virtue of what sorts of facts are we sometimes licensed to draw inferences from similarity? But with so many issues already left outstanding, devising an epistemological story that addresses the justification for analogy's inductive use may well appear premature.

In what follows, let us reason through different scenarios. If there is no case to be made for the inductive view and the heuristic view is correct, there is no epistemological story to be found that justifies an inductive use of analogy. This is not to say, however, that no other epistemological questions arise. Among other things, defenders of the heuristic view may still want to provide an account of the methodology that lies behind working applications of analogical reasoning in science: analogue simulations, extrapolations, model transfer, clusterization, and so on. In each case, one would need to determine whether they ground reliable inferences and, if they do, in virtue of what that happens. This may be accomplished in the traditional philosophical fashion, by spelling out ex ante conditions for when inferences of a specific type may confer justification; alternatively, one may refuse such an approach and argue, alongside Norton's (2021) material theory of induction, that whether or not an inductive inference is justified depends on the facts that ground it – and which can only be known upon empirical investigation. The distinguishing trait of the heuristic view is the claim that, in reconstructing the inferential practice used by scientists, similarity turns out to play at most a marginal role: accordingly, the justificatory work in each case is accomplished by something else, if it is accomplished at all. An example is the abovementioned reconstruction of the methodology of analogue experimentation (Section 3.1.2) as aiming purely at the falsification of general theories, rather than at establishing any inductive justification (Plott 1991). We emphasize that, for many of the inferential practices discussed in Section 3, coherent reconstructions that are consistent with the heuristic view are yet to be fully developed.

An alternative scenario is that in which there is a case to be made for the inductive view and the heuristic view is incorrect. In that case, one would face two orders of philosophical problems: in addition to providing a reconstruction of the different inferential practices in which analogical reasoning is involved, one would have to face the further problem of articulating the relation that analogy may bear with achievements such as truth and knowledge. At least on the face of it, it is not obvious that such a relation should exist. What sorts of facts about us or about the world make it the case that, in at least some contexts, our judgments of similarity offer a reliable guide to adequate models, true theories, or correct predictions? To further clarify the issues at stake, two distinct dimensions of the epistemological problem can be treated as separate. The first is a question about the limits of the proposal: in what domain or up to what extent are we justified to draw analogical inferences about the unknown? That is, are there any empirical or a priori limits to the possibility of justification from analogy? The second problem concerns whether the envisioned epistemological story locates the justification in some a priori principle of rationality or, alternatively, in facts about the world or our relation to it that we access a posteriori.

Neither of these problems has an easy resolution. With regard to the first problem, let us consider the prospects for defending a version of the inductive view when considering the specific case of physics. Even a cursory look at the practice of the past century (or so) of physical science suggests that scientists have often formed, through analogies with more familiar phenomena, expectations about domains that thus far escaped the possibilities of empirical testing – from the behavior of very distant stars to that of bosons. Among other things, this observation is at the basis of Steiner's (1998) *applicability problem of mathematics* to the empirical science – a close relative of what physicist Eugene Wigner once described as the "unreasonable effectiveness of mathematics in the empirical sciences" (cf. also Achinstein 2013). But where do the possibilities of justification from analogy extend to? Some philosophers have argued that justification from analogy can concern at most hypotheses that are in principle testable (Crowther et al. 2019). It would follow from such a view that not all types of limitations to empirical access that we have listed in Section 1.2.1 are on a par from the viewpoint of justification. For instance, hypotheses about very distant stars may be testable but hypotheses about what goes on inside a black hole may not be. For other philosophers, instead, justification is not limited to hypotheses that can be tested. Classic discussions in philosophy about the status of theoretical entities and their alleged distinctiveness from "observables" (see, e.g., Van Fraassen 1980) are relevant to this issue.

The second problem, concerning the envisioned type of justification, is equally thorny. In this context, we can appreciate the problem that struck Hume, viz., that for inductive reasoning we seem unable to provide a justification that is either deductive or inductive. The conceivable alternatives are indeed limited. One approach is to start from the observation that scientists *do* use analogy with an inductive function and insist, with Keynes (1921) and Hesse (1974), that it is a postulate of scientific rationality that there is only a limited number of forms that reality can take. In other words, the fact that *similarity breeds similarity* – that similar effects tend to follow from similar causes, and vice versa – can be related to the assumption that certain properties in nature appear in regular association with others, limiting the number of laws and other patterns that science could expect to discover. This assumption of unity or simplicity in nature is not something that we strictly speaking learn inductively, but is rather to be viewed as a *postulate* for undertaking research in the first place. In other words, deciding to undertake scientific investigation involves a choice to be disposed to believing there is only so much variability in the forms that reality can take. Without it, scientific research would simply have no direction or rule, since for any regularity that we may count as a scientific discovery, we would need to expect an infinity of exceptions.

Postulationist justifications may be considered broadly a posteriori, in that they purport to apply to sufficiently complex worlds like ours. However, an a priori approach may also be attempted. In this case, one ought to show that any rational agent, regardless of the world they inhabit, will use analogy inductively. An important proposal of this kind is developed by Bartha (2009), who connects the inductive role of analogy to two allegedly plausible principles of epistemic rationality:

> *Reflection*: My present credence about a hypothesis H must lie in a range spanned by the possible credences I may come to have about H at a later time (cf. Bartha 2009: 294).

> *Symmetry*: My credence about a H must lie in the range spanned by the credence I hold towards other hypotheses in structurally similar "epistemic situations" (cf. Bartha 2009: 298).

Reflection is a centerpiece of Bayesian epistemology (Van Fraassen 1995). Despite facing objections, Bartha (2009) provides a defense of *Symmetry* as an additional rationality principle. A line of reasoning that Bartha (2009) refers to as the "factorization argument" aims to show that an agent who considers an analogical argument from a well-confirmed hypothesis H* to another hypothesis H in a distinct evidentially impoverished context – one to which they initially assign either no or very small credence – ought to assign some non-negligible credence to H upon assessing the analogical argument as plausible. The supposed justification goes through the principles of *Reflection* and *Symmetry*. First, a rational agent who assigns relatively high credence to H* should consider that, by *Reflection,* in a (possibly imaginary) earlier situation in which there was scarce evidence in favor of H*, a non-negligible credence to H* was still rational due to its being currently well-confirmed. Second, because we can construe this epistemic situation toward H* as similar as we need to the one they experience toward H, by *Symmetry* the agent ought to give non-negligible credence to H. On this account, it is the analogy between H's and H*'s respective domains that ensures the conditions of application of *Reflection* and *Symmetry*.

Without taking a stance as to whether Bartha's proposal succeeds, we regard it as a key objective for the inductive view to develop additional options for addressing the problem of justification. One potential avenue appeals to *styles of reasoning* (Hacking 1992). Rather than specifying a metaphysical postulate, such a view would observe that Analogy is a fundamental, historically grounded form of scientific inference, possessing self-standing validity. *Contextual* or *practice-oriented* approaches may be a further option for shifting the focus from external validation to the internal logic of research. For instance, important work in the context of the historical sciences (Wylie 2011; Currie 2018) emphasizes how analogy fulfills an epistemic role within the broader strategy of connecting and stabilizing distinct lines of evidence – including material traces, models, and theoretical frameworks. Meanwhile, a recent proposal by Hartmann (2025) elaborates on a long-suspected link between the use of analogy in physics and the value of unification (cf. Bartha 2009) by noting how physical analogies achieve a form of inter-theoretic reduction – a relation that has many of the features of reduction, except that there is no sense in which the target system "just is" the source. Connecting the insights that derive from these distinct approaches may open up new pathways for addressing the general problem of justification.

References

Achinstein, P. 1964. Models, analogies, and theories. *Philosophy of Science* 31: 328–349.

Achinstein, P. 2013. *Evidence and Method: Scientific Strategies of Isaac Newton and James Clerk Maxwell*. New York: Oxford University Press.

Agassi, J. 1964. Discussion: Analogies as generalizations. *Philosophy of Science* 31: 351–56.

Alvarez L. W., Alvarez W., and Asaro F. 1980. Extraterrestrial cause for the Cretaceous–Tertiary extinction. *Science* 208(4448): 1095–1108.

Agassi, J. 1988. Analogies hard and soft. In Helman, D. (ed.), *Analogical Reasoning: Perspectives of Artificial Intelligence, Cognitive Science, and Philosophy*. Dordrecht: Kluwer Academic Publishers, 401–419.

Ankeny, R. A. and Leonelli, S. 2020. *Model Organisms*. Cambridge: Cambridge University Press.

Arioli, G. and Valente, G. 2021. What is really quantum in quantum econophysics? *Philosophy of Science* 88(4): 665–685.

Azzouni, J. 2022. Syntactically recharacterizing analogies, assessing theories of assessing analogies (and making some observations about induction too). In Wuppuluri, S., and Grayling, A. (eds.), *Metaphors and Analogies in Sciences and Humanities*. Cham: Springer, 3–32.

Baaquie, B. E. 2004. *Quantum Finance: Path Integrals and Hamiltonians for Options and Interest Rates*. Cambridge: Cambridge University Press.

Barceló, C. 2019. Analogue black-hole horizons. *Nature Physics* 15: 210–213.

Bartha, P. 2009. *By Parallel Reasoning*. New York: Oxford University Press.

Bartha, P. 2022. Prospects for analogue confirmation. *Philosophy of Science* 89(5): 928–938.

Bailer-Jones, D. M. 2002. Models, metaphors and analogies. In Machamer P. K. and Silberstein M. (eds.), *The Blackwell Guide to the Philosophy of Science*. Malden, MA: Blackwell, 108–127.

Bekenstein J. 1972. Black holes and the second law. *Lettere al Nuovo Cimento*, 4: 737–740.

Berieter C. 1985. Towards a solution of the learning paradox. *Review of Educational Research*. 55: 201–206.

Bird, A. 2008. Incommensurability naturalized. In Soler L., Sankey H., and Hoyningen-Huene P. (eds.), *Rethinking Scientific Change and Theory Comparison: Stabilities, Ruptures, Incommensurabilities*. Dordrecht: Springer, 21–39.

Bohr, N. 1913. I. On the constitution of atoms and molecules. *The London, Edinburgh, and Dublin Philosophical Magazine and Journal of Science* 26(151): 1–25.

Bokulich, A. 2015. Maxwell, Helmholtz, and the unreasonable effectiveness of the method of physical analogy. *Studies in History and Philosophy of Science* 50: 28–37.

Boniolo F., Boniolo G., and Valente G. 2023a. Prediction via similarity: Biomedical big data and the case of cancer models. *Philosophy & Technology* 36(1): 8.

Boniolo G., Campaner R., and Carrara M. 2023b. Patient similarity in the era of precision medicine: A philosophical analysis. *Erkenntnis* 88(7): 2911–2932.

Bradley, S. and Thebault, K. P. Y. 2019. Models on the move: Migration and imperialism. *Studies in the History and Philosophy of Science Part A* 77: 81–92.

Campbell, N. R. 1920. *Physics: The Elements*. Cambridge: Cambridge University Press.

Cangiotti, N. and Nappo, F. 2023. Reasoning by analogy in mathematical practice. *Philosophia Mathematica* 31(2): 176–215.

Carnap, R. 1980. A basic system of inductive logic, part II. In Jeffrey, R. (ed.), *Studies in Inductive Logic and Probability*, vol. 2. Berkeley, CA: University of California Press, 7–156.

Castellani, E. 2024. Convergence strategies for theory assessment. *Studies in History and Philosophy of Science* 104: 78–87.

Chen, R. 2020. Natural analogy: A Hessean approach to analogical reasoning in theorizing. *Erkenntnis* 87(5): 2285–2306.

Colhoun, J. and Gentner, D. 2009. Inference process in causal analogies. In Kokinov, B., Holyoak, K., and Gentner, D. (eds.), *Proceedings of the Second International Conference on Analogy*. Sofia: NBU Press, 82–96.

Crowther, K., Linnemann, N. S., and Christian Wutrich, C. 2019. What we cannot learn from analogue experiments. *Synthese* 198 *(Suppl 16)*: 1–26.

Currie, A. 2013. Convergence as evidence. *British Journal for Philosophy of Science*. 64(4): 763–786.

Currie, A. 2018. *Rock, Bone, and Ruin*. Cambridge, MA: MIT Press.

Currie, A. and Killin, A. 2019. From things to thinking: Cognitive archaeology. *Mind & Language*. 34(2): 263–279.

Dardashti, R., Thébault, K., and Winsberg, E. 2017. Confirmation via analogue simulation: What dumb holes could tell us about gravity. *British Journal for Philosophy of Science* 68(1): 55–89.

Dardashti, R., Hartmann, S., Thébault, K., and Winsberg, E. 2019. Hawking radiation and analogue experiments: A Bayesian analysis. *Studies in History and Philosophy of Science (Part B)* 67: 1–19.

Douven, I., Verheyen, S., Elqayam, S., Gärdenfors, P., and Osta-Vélez, M. 2025. Analogical reasoning: A Carnapian approach. *Synthese* 205(5): 1–25.

Dunbar, K. and Blanchette, I. 2001. The in vivo/in vitro approach to cognition: The case of analogy. *Trends in Cognitive Science* 5(8): 334–339.

Falkenhainer, B., Forbus, K. D., and Gentner, D. 1989. The structure-mapping engine: Algorithm and examples. *Artificial Intelligence* 41: 1–63.

Forbus, K.D., Gentner, D., and Law, K. 1995. MAC/FAC: A model of similarity-based retrieval. *Cognitive Science* 19: 141–205.

Fraser, D. 2020. The development of renormalization group methods for particle physics: Formal analogies between classical mechanics and quantum field theories. *Synthese* 197(7): 3027–2063.

Fraser, D. and Koberiski, A. 2016. The Higgs mechanism and superconductivity: A case study of formal analogies. *Studies in History and Philosophy of Modern Physics* 55: 72–91.

Fraser, D. 2022. Justifying the use of purely formal analogies in physics. Unpublished manuscript. PhilSci Archive: 20494.

Frigg, R. and Nguyen, J. 2020. *Modelling Nature: An Opinionated Introduction to Scientific Representation*. New York: Oxford University Press.

Galilei, G. 1610. The Starry Messenger in *Discoveries and Opinions of Galileo*. Translated by Stillman Drake. Garden City, New York: Doubleday Anchor, 1957, 27–58.

Gärdenfors, P. 2000. *Conceptual Spaces: The Geometry of Thought*. Cambridge, MA: MIT Press.

Genta, B. 2020. How to think about analogical inferences: A reply to Norton. *Studies in History and Philosophy of Science (Part A)* 82: 17–24.

Gentner, D. 1983. Structure-mapping: A theoretical framework for analogy. *Cognitive Science* 7(2): 155–170.

Gentner, D. 1989. The mechanisms of analogical learning. In Vosniadou, S. and Ortony, A. (eds.), *Similarity and Analogical Reasoning*. London: Cambridge University Press, 199–241.

Gentner, D. 2010. Bootstrapping the mind: Analogical processes and symbol systems. *Cognitive Science* 34:752–775.

Gentner, D. and Medina, J. 1998. Similarity and the development of rules. *Cognition* 65: 263–297.

Gentner, D. and Forbus, K. 2011. Computational models of analogy. *Wylie Interdisciplinary Reviews: Cognitive Science* 2(3): 266–276.

Gibson, T. 2025. Conceptual revision: How Darwin's analogy supported his theory. *Biology & Philosophy* 40(16): 1–25.

Giere, R. N. 2004. How models are used to represent reality. *Philosophy of Science*, 71: 742–752.

Goodman, N. 1955. *Fact, Fiction, and Forecast*. Cambridge, MA: Harvard University Press.

Goodman, N. 1972. Seven strictures on similarity. In *Problems and Projects*. Indianapolis/ New York: Bobbs-Merrill, 437–446.

Grenier, P., Parent, A., Huard, D., Anctil, F., and Chaumont, D. 2013. An assessment of six dissimilarity metrics for climate analogues. *Journal of Applied Meteorology and Climatology* 52(4): 733–752.

Grüne-Yanoff, T. 2009. Learning from minimal economic models. *Erkenntnis* 70(1): 81–99.

Guala, F. 2005. *The Methodology of Experimental Economics*. Cambridge: Cambridge University Press.

Guala, F. 2010. Extrapolation, analogy, and comparative process tracing. *Philosophy of Science* 77(5): 1070–1082.

Hacking, I. 1992. "Style" for historians and philosophers. *Studies in History and Philosophy of Science* 23(1): 1–20.

Hallegatte, S., Hourcade, J., and Ambrosi, P. 2007. Using climate analogues for assessing climate economic impacts in urban areas. *Climatic Change* 82: 47–60.

Hartmann, S. 2025. Analogy-based reduction: A non-deductive model of intertheoretic integration. Unpublished manuscript.

Hawking, S. 1976. Particle creation by black holes. *Communications in Mathematical Physics* 43: 199–210.

Hempel, C. 1965. *Aspects of Scientific Explanation and Other Essays in the Philosophy of Science*. New York: Free Press.

Herfeld, C. 2024. Model transfer in science. In Knuuttila, T., Carrillo, N., and Koskinen, R. (eds.), *The Routledge Handbook of Scientific Modeling*. New York: Routledge, 270–283.

Herfeld, C. 2025. Model transfer in the recent history of economics: The case of rational choice models. *Perspectives on Science* 33(5): 605–651.

Hesse, M. 1966. *Models and Analogies in Science*. Notre Dame: Notre Dame University Press.

Hesse, M. 1974. *The Structure of Scientific Inference*. London: Macmillan.

Higgs, P. W. 1964. Broken symmetries and the masses of Gauge Bosons. *Physical Review Let*ters 13: 508.

Holyoak, K. 2025. *The Human Edge: Analogy and the Roots of Creative Intelligence*. Cambridge, MA: MIT Press.

Holyoak, K. and Koh, K. 1987. "Surface and structural similarity in analogical transfer". Memory and Cognition, 15(4): 332–340.

Holyoak, K. and Thagard, P. 1989. Analogical mapping by constraint satisfaction. *Cognitive Science* 13(3): 295–355.

Holyoak, K. and Thagard, P. 1995. *Mental Leaps*. Cambridge, MA: MIT Press.

Humphreys, P. 2002. Computational models. *Philosophy of Science* 69(3): 1–11.

Humphreys, P. 2004. *Extending Ourselves: Computational Science, Empiricism, and Scientific Method*. Oxford: Oxford University Press.

Humphreys, P. 2019. Knowledge transfer across scientific disciplines. *Studies in History and Philosophy of Science Part A* 77: 112–19.

Johansen, A., Ledoit, O., and Sornette, D. 2000. Crashes as critical points. *International Journal of Theoretical and Applied Finance* 03(02): 219–55.

Jhun, J., Palacios, P., and Weatherall, J. O. 2018. Market crashes as critical phenomena? Explanation, idealization, and universality in econophysics. *Synthese* 195 (10): 4477–4505.

Kagel, J. H. and Levin, D. 1986. The winner's curse and public information in common value auctions. *The American Economic Review* 76(5): 894–920.

Keynes, J. M. 1921. *A Treatise on Probability*. London: Macmillan.

Kitcher, P. 1993. *The Advancement of Science: Science without Legend, Objectivity without Illusions*. New York: Oxford University Press.

Knuuttila, T. and Loettgers, A. 2014a. Magnets, spins, and neurons: The dissemination of model templates across disciplines. *The Monist* 97(3): 280–300.

Knuuttila, T. and Loettgers, A. 2014b. Varieties of noise: Analogical reasoning in synthetic biology. *Studies in History and Philosophy of Science A* 48: 76–88.

Knuuttila, T. and Loettgers, A. 2016. Model templates within and between disciplines: From magnets to gases – and socio-economic systems. *European Journal for Philosophy of Science* 6(3): 377–400.

Knuttila, T. and Loettgers, A. 2019. Magnetized memories: Analogies and templates in model transfer. In Holm S. and Serban M. (eds.), *Biology: Living Machines?* New York: Routledge, 123–140.

Kuhn, T. 1962. *The Structure of Scientific Revolutions*. Chicago: University of Chicago Press.

LaFollette, H. and Shanks, N. 1993. Animal models in biomedical research: Some epistemological worries. *Public Affairs Quarterly* 7(2): 113–130.

Lange, M. 2016. *Because without Cause*. New York: Oxford University Press.

Lee. H. and Holyoak, K. 2008. The role of causal models in analogical inference. *Journal of Experimental Psychology: Learning, Memory, and Cognition* 34: 1111–2.

Levy, A. and Currie, A. 2015. Model organisms are not (theoretical) models. *British Journal for the Philosophy of Science* 66(2): 327–348.

Lipton, P. 2001. What good is an explanation? In Hon, G. and Rakover, S. (eds.), *Explanation: Theoretical Approaches and Applications*. Dordrecht: Springer Netherlands, 43–59.

Macagno, F., Walton, D., and Tindale, C. 2017. Analogical arguments: Inferential structures and defeasibility conditions. *Argumentation* 31(2): 221–243.

Mach, E. 1905. *Erkenntnis und Irrtum: Skizzen zur Psychologie der Forschung*. Leipzig: Johann Ambrosius Barth.

Mäki, U. and Marchionni, C. 2011. Is geographical economics imperializing economic geography? *Journal of Economic Geography* 11(4): 645–665.

Markman, A. and Gentner, D. 1993. Structural alignment during similarity comparisons. *Cognitive Psychology* 25(4): 431–467.

Maxwell, J. C. 1890. *The scientific papers of James Clerk Maxwell*. Niven, W. D. (ed.) Cambridge: Cambridge University Press.

Mitchell, M. 1993. *Analogy-Making as Perception: A Computer Model*. Cambridge, MA: MIT Press.

Mitchell, M. 2021. Abstraction and analogy-making in artificial intelligence. *Annals of the New York Academy of Sciences* 1505: 79–101.

Nappo, F. 2020. Learning from non-causal models. *Erkenntnis* 87(1): 2419–39.

Nappo, F. 2021. The double nature of Maxwell's physical analogies. *Studies in History and Philosophy of Science* 89: 212–225.

Nappo, F. 2022. Confirmation by analogy. *Synthese* 200(35): 1–22.

Nappo, F., Magli, G., and Valente, G. 2023. Evidence and analogy in Archaeoastronomy. *Synthese* 200(439): 1–25.

Nappo, F. 2025. Hesse on materiality and induction. In Gori P. (ed.), *Mary B. Hesse (1924–2016): Metaphors, Models, and the Growth of Scientific Knowledge*. Cham: Springer, 179–198.

Nappo. F. and Cangiotti, N. 2025. What's so special about black hole simulations? *European Journal for Philosophy of Science* 15(36):1–25.

Nersessian, N. 2008. *Creating Scientific Concepts*. Cambridge, MA: MIT Press.

Nersessian, N. and Chandrasekharan, S. 2009. Hybrid analogies in conceptual innovation in science. *Cognitive Systems Research* 10(3):78–87.

Niss, M. and Blum, W. 2020. *The Learning and Teaching of Mathematical Modeling*. London: Routledge.

Norton, J. 2021. *The Material Theory of Induction*. BSPS Open Book.

Nyrup, R. 2020. Of water drops and atomic nuclei: Analogy and pursuit-worthiness in science. *The British Journal for the Philosophy of Science* 71(3): 881–903.

Nyrup, R. 2021. Three uses of analogy: A philosophical view of the archaeologist's toolbox. *Interarchaeologia 6: Archaeology and Analogy*. Helsinki: University of Helsinki.

Oppenheimer R. 1956. Analogy in science. *American Psychologist* 11(3): 127–135.

Parker, W. S. 2010. Comparative process tracing and climate change fingerprints. *Philosophy of Science* 77(5): 1083–1095.

Palacios, P. and Jhun J. 2023. Statistical mechanical models of finance. In Knuuttila, T., Carrillo, N., and Koskinen, R. (eds.), *The Routledge Handbook of Philosophy of Scientific Modeling*. London: Routledge, 429–442.

Pietsch, W. 2021. *Big Data*. Cambridge: Cambridge University Press.

Plott, C. 1991. Will economics become an experimental science? *Southern Economic Journal* 57(4) 901–919.

Polya, G. 1954. *Mathematics and Plausible Reasoning. Vol I: Induction and Analogy in Mathematics*. Princeton: Princeton University Press.

Polya, G. 1957. *How to Solve It: A New Aspect of Mathematical Method*. Princeton: Princeton University Press.

Popper, K. 1959. *Logic of Scientific Discovery*. London: Hutchinson.

Richland, L., Morrison, G., and Holyoak, K. 2006. Children's development of analogical reasoning: Insights from scene analogy problems. *Journal of Experimental Child Psychology* 94: 249–271.

Rutherford, E. 1911. The scattering of α and β particles by matter and the structure of the atom. *Philosophical Magazine Series 6* 21(125): 669–688.

Shelley, C. 2002. Analogy counterarguments and the acceptability of analogical hypotheses. *The British Journal for the Philosophy of Science* 53: 477–492.

Schelling, T. C. 1969. Models of segregation. *The American Economic Review* 59(2): 488–493.

Sklar, L. 2001. *Theory and Truth: Philosophical Critique within Foundational Science*. New York: Oxford University Press.

Skyrms, B. 1993. Analogy by similarity in hyper-Carnapian inductive logics. In Earman, J. (ed.), *Philosophical Problems of the Internal and External Worlds*. Pittsburgh: University of Pittsburgh Press, 273–282.

Steel, D. 2008. *Across the Boundaries: Extrapolation in Biology and Social Sciences*. Oxford: Oxford University Press.

Steel, D. 2010. A new approach to argument by analogy: Extrapolation and chain graphs. *Philosophy of Science* 77(5): 1058–1069.

Stegenga, J. 2024. Fast science. *British Journal for the Philosophy of Science*. Forthcoming.

Steiner, M. 1998. *The Applicability of Mathematics as a Philosophical Problem*. Cambridge, MA: Harvard University Press.

Sterrett, S., 2017, "Experimentation on analogue models", In Magnani, L., Bertolotti, T. (eds.) Springer Handbook of Model-Based Science. pp. 857–878. Cham: Springer.

Sterrett, S. 2020. Scale modeling. In Michelfelder, D. and Doorn, E. (eds.), *The Routledge Handbook of Philosophy of Engineering*. New York: Routledge, 394–407.

Suárez, M. 2003. Scientific representation: Against similarity and isomorphism. *International Studies in the Philosophy of Science* 17(3): 225–244.

Suárez, M. 2024. *Inference and Representation: A Study in Modelling Science*. Chicago: University of Chicago Press.

Sugden, R. 2009. Credible worlds, capacities and mechanisms. *Erkenntnis* 70(1): 3–27.

Suppes, P. 1962. Models of data. In Nagel E., Suppes P., and Tarski A. (eds.), *Logic, Methodology and Philosophy of Science: Proceedings of the 1960 International Congress*. Palo Alto: Stanford University Press, 252–261.

Swoyer, C. (1991). Structural representation and surrogative reasoning. *Synthese* 87(3): 449–508.

Unruh, W. 1981. Experimental black-hole evaporation? *Physical Review Letters* 46(21): 1351–53.

Unruh, W. G. and Schutzhold, R. 2005. Universality of the Hawking effect. *Physical Review D* 71(2): 024028.

Valente, G., Bobadilla, H., El Skaf, R., and Nappo, F. 2024. Tales of twin cities: What are climate analogues good for? *European Journal for Philosophy of Science* 14(3): 1–28.

van Eersel, G. G., Koppenol-Gonzalez, G. V., and Reiss, J. 2019. Extrapolation of experimental results through analogical reasoning from latent classes. *Philosophy of Science* 86(2): 219–235.

Van Fraassen, B., 1995. Belief and the problem of Ulysses and the syrens. Philosophical Studies. 77(1): 7–37.

Walton, D., Reed, C., and Macagno, F. 2012. *Argumentation Schemes*. Cambridge: Cambridge University Press.

Weisberg, M. 2013. *Simulation and Similarity: Using Models to Understand the World*. New York: Oxford University Press.

Wylie, A. 1985. The reaction against analogy. *Advances in Archaeological Method and Theory* 8: 63–111.

Wylie, A. 2011. Critical distance: stabilizing evidential claims in archaeology. In Dawid, P. Twining, W. and Vasilaki, M. (eds.) Evidence, Inference, and Enquiry. New York: Oxford University Press. 371–394.

Zuchowski, L. 2019. Modeling and knowledge transfer in complexity science. *Studies in History and Philosophy of Science Part A* 77: 120–129.

Acknowledgments

The authors thank Paul Bartha, Nicolò Cangiotti, Adrian Currie, Francesco Guala, Stephan Hartmann, Catherine Herfeld, Tarja Knuuttila, and Patricia Palacios for their comments. Also, a special thanks goes to the editor Jacob Stegenga and to two anonymous referees. G. Valente was supported by the MIUR - PRIN 2022 project "Analogical Reasoning in Contemporary Physical Theories" (Prot. N. 2022F4Z8YH).

Cambridge Elements

Philosophy of Science

Jacob Stegenga

NTU Singapore

Jacob Stegenga is a Professor at NTU Singapore, and previously taught at the University of Cambridge. He has published widely in philosophy of science and philosophy of medicine, and is the author of *Medical Nihilism*, described as 'a landmark work', *Care and Cure: An Introduction to Philosophy of Medicine*, and a book to be published in 2025 titled *Heart of Science*.

About the Series

This series of Elements in Philosophy of Science provides an extensive overview of the themes, topics and debates which constitute the philosophy of science. Distinguished specialists provide an up-to-date summary of the results of current research on their topics, as well as offering their own take on those topics and drawing original conclusions.

Cambridge Elements Ξ

Philosophy of Science

Elements in the Series

Scientific Models and Decision Making
Eric Winsberg and Stephanie Harvard

Science and the Public
Angela Potochnik

Feminist Philosophy of Science
Anke Bueter

Abductive Reasoning in Science
Finnur Dellsén

Climate Science
Wendy S. Parker

The Social Dimensions of Scientific Knowledge: Consensus, Controversy, and Coproduction
Boaz Miller

Scientific Realism
Timothy D. Lyons

Science, Pseudoscience, and the Demarcation Problem
Dániel Bárdos and Adam Tamas Tuboly

Underdetermination and Theoretical Virtues
Dana Tulodziecki

The Philosophy of Linguistics
Ryan M. Nefdt

Science Denial: Post-Truth or Post-Trust?
Gabriele Contessa

Analogical Reasoning in Science
Francesco Nappo and Giovanni Valente

A full series listing is available at: www.cambridge.org/EPSC

For EU product safety concerns, contact us at Calle de José Abascal, 56–1°,
28003 Madrid, Spain or eugpsr@cambridge.org.

www.ingramcontent.com/pod-product-compliance
Ingram Content Group UK Ltd.
Pitfield, Milton Keynes, MK11 3LW, UK
UKHW020843290926
15470UKWH00004B/1514
* 9 7 8 1 0 0 9 5 2 6 4 6 3 *